D1538651

Proceedings of the 1st Asia-Pacific Conference

QUANTUM
INFORMATION SCIENCE

Proceedings of the 1st Asia-Pacific Conference

QUANTUM
INFORMATION SCIENCE

National Cheng Kung University, Taiwan
Dec 10–13, 2004

editors

Chopin Soo • Wei-Min Zhang
National Cheng Kung University, Taiwan

 **World Scientific**

NEW JERSEY • LONDON • SINGAPORE • BEIJING • SHANGHAI • HONG KONG • TAIPEI • CHENNAI

Published by

World Scientific Publishing Co. Pte. Ltd.

5 Toh Tuck Link, Singapore 596224

USA office: 27 Warren Street, Suite 401-402, Hackensack, NJ 07601

UK office: 57 Shelton Street, Covent Garden, London WC2H 9HE

British Library Cataloguing-in-Publication Data

A catalogue record for this book is available from the British Library.

QUANTUM INFORMATION SCIENCE
Proceedings of the 1st Asia-Pacific Conference

Copyright © 2005 by World Scientific Publishing Co. Pte. Ltd.

All rights reserved. This book, or parts thereof, may not be reproduced in any form or by any means, electronic or mechanical, including photocopying, recording or any information storage and retrieval system now known or to be invented, without written permission from the Publisher.

For photocopying of material in this volume, please pay a copying fee through the Copyright Clearance Center, Inc., 222 Rosewood Drive, Danvers, MA 01923, USA. In this case permission to photocopy is not required from the publisher.

ISBN 981-256-460-8

Printed in Singapore by Mainland Press

PREFACE

The goals of the Conference were to promote and strengthen the interactions and exchange of knowledge among researchers of the Asia-Pacific region in the rapidly advancing field of Quantum Information Science. We also hope that the Asia-Pacific Conference on Quantum Information Science reflects a cooperative effort, and that the first gathering in Tainan on December 10th.-13th. 2004 marks the beginning of a series of regular meetings on Quantum Information Science which will be held and hosted by different institutions in the Asia-Pacific region.

We would like to once again express our sincere appreciation to all speakers and participants of the 1st Asia-Pacific Conference on Quantum Information Science. We wish also to thank all members of our advisory committee, and all international coordinators and local organizers for their generous assistance; and our colleagues for their kind support. It gives us great pleasure and satisfaction to record our gratitude to the conference administrative assistants and to all our student volunteers for their invaluable help.

Chopin Soo (Conference Chairman)
Wei-Min Zhang (Co-chairman)

Department of Physics,
and Center for Quantum Information Science,
National Cheng Kung University,
Tainan, Taiwan.

Group photo of the 1st Asia-Pacific Conference on Quantum Information Science
Main Library, National Cheng Kung University, Tainan, Taiwan

SPONSORS

National Center for Theoretical Sciences
Center for Quantum Information Science, National Cheng Kung University
Micro-Nano Technology Research Center, National Cheng Kung University
Physics Research Promotion Center, National Science Council, R.O.C.
Ministry of Education, R.O.C.

COMMITTEES

International Advisory Committee

Artur Ekert	Cambridge Univeristy, United Kingdom, and Nat. University of Singapore
Guangcan Guo	University of Science and Technology, P. R. China
Ting-Kuo Lee	Academia Sinica, and Nat. Sc. and Tech. Program for NanoSc. and Nanotech., Taiwan
Gerard J. Milburn	University of Queensland, Australia
Lu Jue Sham	University of California San Diego, USA
Jaw-Shen Tsai	NEC, Japan
Maw-Kuen Wu	Academia Sinica, and National Science Council, Taiwan

International Coordinators/Organizers

Hsi-Sheng Goan	U. New South Wales, Australia; and Nat. Taiwan U., Taiwan
L. C. Kwek	Nat. Inst. of Education, Singapore
Soonchil Lee	KAIST, Rep. Korea
C. H. Oh	Nat. U., Singapore
K. K. Phua	World Scientific Publishing Co; and Nat. U. of Singapore
Chopin Soo	Nat. Cheng Kung U., Taiwan (Conference Chairman)
Xiang-bin Wang	ERATO, Japan
Wei-Min Zhang	Nat. Cheng Kung U., Taiwan (Co-chairman)

Local Organizers

Chia-chu Chen	Nat. Cheng Kung U.
Ching Cheng	Nat. Cheng Kung U.
Chi-Yee Cheung	Academia Sinica
Chien-er Lee	Nat. Cheng Kung U.
Chuan-Pu Liu	Nat. Cheng Kung U.
Yan-Ten Lu	Nat. Cheng Kung U.
Zheng-Yao Su	Nat. Cen. for High Performance Computing
Chin-Chun Tsai	Nat. Cheng Kung U.
Yin-Zhong Wu	Nat. Cheng Kung U., and Changshu Inst. of Tech.

Conference administrative assistants

Annie Yi-Zhen Lin	Nat. Cen. for Theoretical Sciences
Betty Miao-Li Tsai	Cen. for Quantum Information Sc., Nat. Cheng Kung U.

CONTENTS

REALISTIC SIMULATIONS OF SINGLE-SPIN MEASUREMENT VIA MAGNETIC RESONANCE FORCE MICROSCOPY*

TODD A. BRUN

Communication Sciences Institute, University of Southern California,
Los Angeles, CA 90089-2565 USA

HSI-SHENG GOAN

Department of Physics, National Taiwan University,
No. 1, Sc 4, Roosevelt Road, Taipei 106, Taiwan (ROC)

The problem of measuring single electron or nuclear spins is of great interest for a variety of purposes, from imaging the structure of molecules to quantum information processing. One of the most promising techniques is magnetic resonance force microscopy (MRFM), in which the force between a spin and a small permanent magnet resonantly drives the oscillations of a microcantilever. Numerous issues arise in understanding this system: thermal noise in the cantilever, shot-noise and back-action from monitoring the cantilever's motion, spin relaxation, and interaction with higher cantilever modes. Detailed models of these effects allow one to assess their relative importance and the necessary improvements for sensitivity at the single-spin level.

1. Single spin measurement

Single-spin measurement is an extremely important and difficult challenge for modern quantum technology. It is necessary for the success of several spin-based proposals for quantum information processing[1,2,3,4,5]. Single spin detection can be done either *indirectly*, by turning the spin measurement into the movement of a charge (using, e.g., a single-electron transistor); or *directly* by detecting the force produced by the spin's magnetic field. For direct measurement, the most promising approach is magnetic resonance force microscopy (MRFM)[6,7,8].

MRFM has recently reached the sensitivity to detect (but not measure) a single electron spin–a feat which AIP Physics News recently called the top story of 2004[9]. We should note, however, that even "direct" detection is really indirect.

*This work was supported by a hewlett-packard fellowship (HSG), the Chooljian membership in Natural Sciences (TAB), and DOE grant no. DE-FG02-90ER40542 (TAB).

The forces exerted by a single spin are far too weak to be detected by macroscopic devices without a lot of clever tricks!

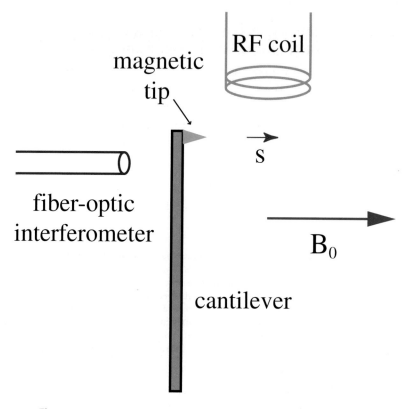

Figure 1.   A schematic diagram of the setup for an MRFM spin measurement.

2. Cyclic adiabatic inversion

For this paper, we will mostly discuss a version of this protocol called *Cyclic Adiabatic Inversion*[8,10,11,12] (CAI). The spin of the electron precesses in a local magnetic field with three components: an imposed external field; the field from a permanent magnet on the tip of a nearby high-Q microcantilever; and the field from incident microwaves.

The microwave field is given a frequency modulation in resonance with the cantilever. The precessing spin moves adiabatically with the field. The electron spin has a dipole-dipole interaction with the permanent magnet, thereby exerting a weak force, which varies periodically in resonance with the cantilever. Over many oscillations, the cantilever is driven by this force until it produces an amplitude which is detectable by an optical interferometer.

The initial orientation of the spin will determine the sign of this driving force. Thus, the output signal is the phase of the resulting cantilever oscillations. These in turn produce an oscillating photocurrent in the optical interferometer. It is this photocurrent which is actually measured. This chain of successive amplifications reminds one of the remarkable inventions of Rube Goldberg. An oscillating spin produces over time an oscillating cantilever, which in turn produces an optical signal, which is finally read out as a photocurrent.

The Hamiltonian which describes the interaction of the spin and the cantilever is

$$\hat{H}_{SZ}(t) = \hat{H}_Z - 2\eta\hat{Z}\hat{S}_z + f(t)\hat{S}_z - \varepsilon\hat{S}_x \,, \tag{1}$$

$$\hat{H}_Z = \frac{1}{2m}\hat{p}^2 + \frac{m\omega_m^2}{2}\hat{Z}^2 \,. \tag{2}$$

Because $f(t)$ and ε are both large compared to the cantilever frequency ω_m, we can switch to a rotating picture and make a rotating wave approximation. In this approximation the Hamiltonian becomes

$$H'_{SZ}(t) = \hat{H}_Z - 2\eta g(t)\hat{Z}\hat{S}'_z \,, \tag{3}$$

where $\hat{S}'_z$ is the transformed spin operator and $g(t) = f(t)/\sqrt{f^2(t) + \varepsilon^2}$ is a new periodic function which is still resonant with the cantilever.

We seen from this Hamiltonian that if the spin is initially in the $\pm 1/2$ eigenstate of $\hat{S}'_z$, it remains in that state at all successive times; and this state determines the sign of the driving force acting on the oscillator.

As Fig. 2 shows, the rotating-wave approximation is quite accurate. The measurement is made by looking at the phase of the cantilever oscillation. We see in Fig. 3 that the components at different phases change quite markedly with time, depending on the orientation of the measured spin.

3. Modeling MRFM

To model the complete MRFM measurement process we need more than just the spin-cantilever Hamiltonian. We need to include three degrees of freedom: the cantilever, the spin, and the optical cavity mode. We need to include *thermal noise and damping* in the cantilever, *cavity loss* in the cavity, and the *effects of monitoring* (i.e., measurement back-action).

Because optical frequencies are much higher than spin precession or cantilever oscillation, and the cavity is highly lossy, the cavity mode degree of freedom can

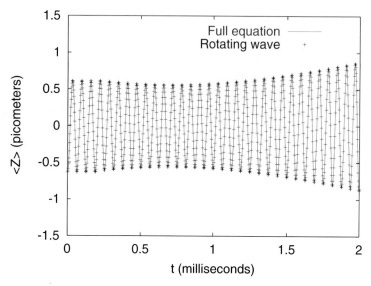

Figure 2. A comparison between the full equations and the rotating-wave approximation for an MRFM system.

be adiabatically eliminated. The result is a *stochastic master equation* for the spin and cantilever degrees of freedom[13,14].

The stochastic master equation is

$$d\rho = -\frac{i}{\hbar}[\hat{H},\rho]dt + \sum_j (2\hat{L}_j\rho\hat{L}_j^\dagger - \{\hat{L}_j^\dagger\hat{L}_j,\rho\})dt$$

$$+\sqrt{2e_d}\left((\hat{L}_2 - \langle\hat{L}_2\rangle)\rho + \rho(\hat{L}_2 - \langle\hat{L}_2\rangle)\right)dW_t . \tag{4}$$

$$M[dW_t] = 0 , \quad M[(dW_t)^2] = dt . \tag{5}$$

The Lindblad operators are

$$\hat{L}_1 = \sqrt{\gamma_m/2}\left((1/\ell)\hat{Z} + i(\ell/\hbar)\hat{p}\right) ,$$

$$\hat{L}_2 = \sqrt{8\kappa^2 E^2/\gamma_c^3}\hat{Z} , \tag{6}$$

and the effective Hamiltonian is

$$\hat{H}(t) = \hat{H}'_{SZ}(t) + \frac{4\kappa E^2}{\gamma_c^2}\hat{Z} + (\gamma_m/2)(\hat{Z}\hat{p} + \hat{p}\hat{Z}) . \tag{7}$$

The $\hat{L}_2$ terms are the result of monitoring, and produce a stochastic localization[15] of the cantilever position Z.

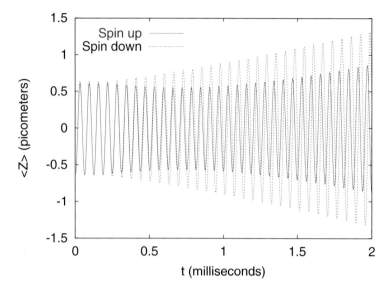

Figure 3. Cantilever position versus time for spin up and spin down.

Because the cantilever position is driven by the sign of the spin, however, this localization in position also produces a localization in the spin $\hat{S}'_z$.

This localization is very fast! In fact, as we can see from Fig. 4, it takes place long before we can actually detect the signal which indicates the spin orientation (since that takes many cantilever oscillations).

4. Stochastic Schrödinger Equation

Solving the stochastic master equation is possible numerically, but in practice is very difficult. It turns out to be far more efficient to *unravel* this master equation into a *stochastic Schrödinger equation* (SSE).

$$d|\psi\rangle = -\frac{i}{\hbar}\hat{H}(t)|\psi_{SZ}\rangle dt + \sum_{j=1}^{2}\left(2\langle\hat{L}_j^\dagger\rangle\hat{L}_j - \hat{L}_j^\dagger\hat{L}_j - |\langle\hat{L}_j\rangle|^2\right)|\psi_{SZ}\rangle dt$$
$$+\sqrt{2}\left(\hat{L}_1 - \langle\hat{L}_1\rangle\right)|\psi_{SZ}\rangle dW_{1t} + \sqrt{2e_d}\left(\hat{L}_2 - \langle\hat{L}_2\rangle\right)|\psi_{SZ}\rangle dW_{2t}$$
$$+\sqrt{2(1-e_d)}\left(\hat{L}_2 - \langle\hat{L}_2\rangle\right)|\psi_{SZ}\rangle dW_{3t} . \tag{8}$$

By averaging over W_1 and W_3 we can reproduce the solution to the stochastic master equation. The advantage of an SSE is that we can take advantage of *localizing effects* which tend to make the solutions remain small wavepackets at all times.

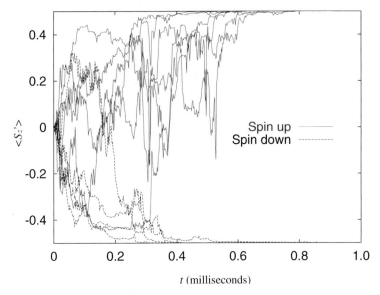

Figure 4.   The expectation value of spin $\langle \hat{S}_z^i \rangle$ as a function of time for ten different trajectories, where the initial spin is in the superposition $(|\uparrow\rangle + |\downarrow\rangle)/\sqrt{2}$. Note that the spins randomly settle on either the state up or down in a short time.

In fact, we might use this tendency to further simplify our description. Suppose that we assume that the state of the cantilever plus spin is

$$|\psi\rangle = \sqrt{r}|\phi_d\rangle|\downarrow\rangle + \sqrt{1-r}|\phi_u\rangle|\uparrow\rangle \,, \tag{9}$$

and $|\phi_d\rangle$ and $|\phi_u\rangle$ remain small Gaussian wavepackets at all times. What kind of description does that give us? We can find a set of coupled differential equations for the wave packet positions and momenta and the spin probability r.

These equations look like this:

$$dZ_d = (p_d/m)dt + \text{extra terms} \,,$$
$$dp_d = (-m\omega_m^2 Z_d - \eta g(t) - 2\gamma_m p_d)dt + \text{extra terms} \,,$$
$$dZ_u = (p_u/m)dt + \text{extra terms} \,,$$
$$dp_u = (-m\omega_m^2 Z_u + \eta g(t) - 2\gamma_m p_u)dt + \text{extra terms} \,,$$
$$dr = 2A(Z_u - Z_d)r(1-r)dW_t. \tag{10}$$

The important point is that the leading terms of these equations are just like the classical equations for damped driven oscillators, and $r \to 0, 1$ at long times. Because these equations mirror the classical equations, it turns out that estimates of this system's behavior based on classical intuition are actually surprisingly accurate. The two curves are plotted in Fig. 5 for the same realization of the noise, and are extremely close.

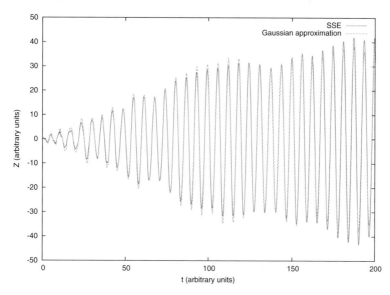

Figure 5. A comparison between the fully quantum equations and the Gaussian approximation for an MRFM system.

5. Spin relaxation and quantum jumps

We have considered several sources of error on the cantilever. To accurately model the real physical system, however, we must also include noise which affects the *spin*. The simplest type we might consider is *spin relaxation*, in which we add to our stochastic master equation terms corresponding to a Lindblad operator of the form

$$\hat{L}_S = \sqrt{1/t_1}\,\hat{S}'_x. \qquad (11)$$

This represents noise in the local environment of the spin which tends to flip its orientation. To be a reasonable measurement technique, our MRFM setup must be able to measure the spin in a time short compared to t_1.

At present we are not very close to that limit. In the recent single-spin detection experiment, the signal was obtained by averaging data over 13 *hours*. By contrast, the spin relaxation time was on the order of hundreds of milliseconds.

However, improving sensitivity by one or two orders of magnitude would make a dramatic difference in the time needed to produce a signal. At that point, true single-spin measurements will be possible.

We can unravel the effects due to spin relaxation as well, to recover a new SSE. In this case, it is convenient to use a different kind of unraveling, into *quantum jumps*. The new SSE contains a term

$$d|\psi\rangle = \cdots + \left(\hat{S}_x' - 1\right)|\psi\rangle dN, \tag{12}$$

where $dN^2 = dN$ and $M[dN] = dt/t_1$. This term is usually zero; but randomly, with rate $1/t_1$, quantum jumps occur. In a jump, $dN = 1$ and the spin orientation is flipped. This SSE is a hybrid form of stochastic equation, containing both diffusive and jump terms.

In the Gaussian approximation, this effect is easily incorporated by having a rate for jumps which switch the "up" and "down" wavepackets, causing $Z_d, p_d \leftrightarrow Z_u, p_u$ and $r \rightarrow 1 - r$.

6. What's left?

At present we include many important effects in our model: thermal noise, measurement back-action, shot-noise and spin-relaxation. What remains to be done?

(1) Actual spin-relaxation need not take the simple form described above; simple noise in the lab frame can become time dependent in the rotating frame.
(2) Recent experiments use a different protocol than CAI, called OSCAR. Our model must be adapted to this different technique.
(3) Interaction between the spin and higher modes of the cantilever is a significant source of noise. This should be included.
(4) In the longer term, we'd like to treat systems with multiple nearby spins.

In general, the goal would be to develop models of sufficient accuracy to help experimenters in assessing the relative merits of different protocols, and perhaps to find the most useful areas for improvement. There is a large gap, at present, between what is theoretically possible and what is experimentally feasible; an important need is to develop the theoretical and numerical tools to help fill that gap.

References

1. D. Loss and D. P. DiVincenzo, Phys. Rev. A **57**, 120 (1998).
2. B.E. Kane, Nature **393**, 133-137 (1998).

3. R. Vrijen, E. Yablonovitch, K. Wang, H. W. Jiang, A. Balandin, V. Roychowdhury, T. Mor, and D. DiVincenzo Phys. Rev. A **62**, 012306 (2000).
4. G.P. Berman, G.D. Doolen, P.C. Hammel, and V.I. Tsifrinovich, Phys. Rev. B **61**, 14694 (2000).
5. J. Twamley, quant-ph/0210202.
6. J.A. Sidles, Appl. Phys. Lett. **58**, 2854 (1991).
7. J.A. Sidles, Phys. Rev. Lett. **68**, 1124 (1992).
8. G.P Berman, F. Borgonovi, G. Chapline, S.A. Gurvitz, P.C. Hammel, D.V. Pelekhov, A. Suter and V.I. Tsifrinovich, J. Phys. A **36**, 4417 (2003).
9. D. Rugar R. Budakian, H. J. Mamin and B. W. Chui, Nature **430**, 329 (15 July 2004).
10. D. Rugar, O. Züger, S. Hoen, C.S. Yannoni, H.M. Vieth, and R.D. Kendrick, Science **264**, 1560 (1994).
11. K. Wago, D. Botkin, C.S. Yannoni, and D. Rugar, Phys. Rev. B **57**, 1108 (1998).
12. T.A. Brun and H.-S. Goan, Phys. Rev. A **68**, 032301 (2003).
13. H. J. Carmichael, *An Open System Approach to Quantum Optics*, Lecture notes in physics (Springer-Verlag, Berlin, 1993).
14. H. M. Wiseman and G.J. Milburn, Phys. Rev. A **47**, 642 (1993); **47**, 1652 (1993).
15. R. Schack, T. A. Brun, I. C. Percival, J. Phys. A. **28**, 5401 (1995).

CONTINUOUS VARIABLE TELEPORTATION OF QUANTUM FIELDS

H. J. CARMICHAEL

Department of Physics, University of Auckland,
Private Bag 92019, Auckland, New Zealand
E-mail: h.carmichael@auckland.ac.nz

A quantum trajectory formulation of broadband continuous variable teleportation is developed. Inputs and outputs are quasi-monochromatic quantum fields rather than single-mode quantum states. The formalism accounts for the continuous measurements of Alice and Victor and continuous displacement of the teleported field by Bob. It is applied to the teleportation of the Mollow spectrum and photon antibunching in single-atom resonance fluorescence.

1. Introduction

Quantum teleportation was introduced by Bennett *et al.*[1] for discrete variable systems, i.e., two-state systems or qubits. The first experimental realization of the protocol was carried out by Bouwmeester *et al.*[2]. Prior to this, Vaidman[3] made the generalization to continuous variables, but in a version unsuitable for experimental implementation. Braunstein and Kimble[4] introduced the necessary modifications for an implementation using two-mode quadrature squeezed light. Experiments on continuous variable teleportation using squeezed light have been carried out by Furusawa *et al.*[5] and Bowen *et al.*[6].

These latter experiments raise some interesting theoretical questions. Teleportation protocols, including the one introduced by Braunstein and Kimble, are executed in a series of distinct steps, on objects characterized by time-independent quantum states. The steps using squeezed light are as follows: (1) preparation of the input state and the two-mode squeezed (entangled) state; (2) dispersal of the input and one entangled mode to Alice, dispersal of the other entangled mode to Bob; (3) execution of two quadrature measurements by Alice; (4) sending of Alice's measurement results to Bob; (5) displacement of the entangled mode held by Bob; and (6) verification by Victor that the resulting output matches the input. This is not the sequence of steps in the experiments,[5,6] however, since for them there *is* no sequence; all six actions are carried out continuously and in parallel. That is not to say that a plausible connection between the experiments and the protocol of Braunstein and Kimble cannot be made. Only that the implementation is a great deal more general than what the existing theory describes. In particular, the continuous and parallel execution of the above list arises from the fact that the experimental implementation is inherently broadband, or more significantly multi-mode. Thus, the implemented teleporter is well able to accept an input as an arbitrary quasi-monochromatic quantum field, rather than a mode, in a quantum state. We are prompted to ask: what quantum field then appears at the output? how do we formulate a multi-mode description capable of addressing this question?

At minimum, a treatment of the continuous measurements of Alice and Victor is needed, together with an accounting for the continuous feed-forward of Alice's measurement results to Bob. Our aim in this paper is to formulate a treatment along these lines using quantum trajectory theory. First, in Sec. 2, we use the quantum trajectory approach to model a pulsed-light implementation of the elementary single-mode teleportation protocol of Braunstein and Kimble.[4] Next, in Sec. 3, we set up the basic equations of a broadband and multi-mode generalization of the model. The developed stochastic Schrödinger equation depends in some of its details on the measurement made by Victor, who continuously monitors the output field. The specific examples of homodyne, heterodyne, and photoelectron counting measurements are considered in Sec. 4, where for the purpose of illustration teleportation of the vacuum field is treated. Section 5 applies the developed formalism to teleportation of the Mollow spectrum and photon antibunching in single-atom resonance fluorescence. We finish with conclusions in Sec. 6.

2. Elementary Single-Mode Proposal

Single-mode continuous variable (CV) teleportation[4] is modeled according to the scheme in Fig. 1. Its execution follows the mentioned steps:

<u>Step 1</u>: Cavity modes A and B are prepared in a two-mode squeezed state, $|\chi_{AB}^{(r)}\rangle$, of finite squeeze parameter r. Mode c is the input, which for definiteness we prepare in the Fock state $|\chi_{\text{in}}\rangle = |2\rangle$, Wigner function as shown in Fig. 2(a).

<u>Step 2</u>: Modes A and c are allowed to leak from their cavities, both with cavity decay rate κ. They are dispersed as exponential pulses to Alice. For simplicity, Bob's mode B remains trapped in the cavity, though nothing changes if it too is dispersed, so long as no measurement on it is made. Fields $\hat{\mathcal{E}}_{\text{in}}$ and $\hat{\mathcal{E}}_{\text{Alice}}$ (Fig. 1) are sums of free fields $\hat{\mathcal{E}}_{\text{in}}^{\text{free}}$ and $\hat{\mathcal{E}}_{\text{Alice}}^{\text{free}}$, and source fields $\sqrt{2\kappa}\hat{c}$ and $\sqrt{2\kappa}\hat{A}$, respectively.

<u>Step 3</u>: After combining $\hat{\mathcal{E}}_{\text{in}}$ and $\hat{\mathcal{E}}_{\text{Alice}}$ at a 50/50 beam splitter, Alice measures

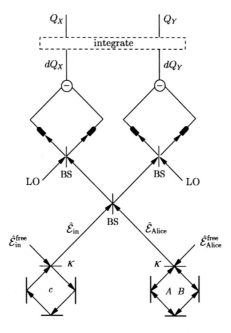

Figure 1.   Schematic of single-mode CV teleportation; modes c, A, and B are resonant.

her dispersed fields using mode-matched homodyne detection (local oscillator $\mathcal{E}_{lo}e^{-\kappa t}$). We model the measurements within quantum trajectory theory. Integrated photocurrents yield electric charges

$$Q_X(t) = \int_0^t e^{-\kappa t'} dQ'_X, \qquad Q_Y(t) = \int_0^t e^{-\kappa t'} dQ'_Y, \qquad (1)$$

where

$$dQ_X = 2\sqrt{\kappa}[\langle \hat{c}_X \rangle_{\text{REC}} + \langle \hat{A}_X \rangle_{\text{REC}}]dt + dW_X, \qquad (2a)$$

$$dQ_Y = 2\sqrt{\kappa}[\langle \hat{c}_Y \rangle_{\text{REC}} - \langle \hat{A}_Y \rangle_{\text{REC}}]dt + dW_Y. \qquad (2b)$$

Here dW_X and dW_Y are independent Wiener increments,

$$\overline{(dW_X)^2} = \overline{(dW_Y)^2} = dt, \qquad \overline{dW_X dW_Y} = 0, \qquad (3)$$

which model shot noise, and the quadrature amplitude expectations, $\langle \hat{c}_{X,Y} \rangle_{\text{REC}}$ and $\langle \hat{A}_{X,Y} \rangle_{\text{REC}}$, are evaluated with respect to the conditional state

$$|\psi_{\text{REC}}\rangle = \frac{|\bar{\bar{\psi}}_{\text{REC}}\rangle}{\sqrt{\langle \bar{\bar{\psi}}_{\text{REC}}|\bar{\bar{\psi}}_{\text{REC}}\rangle}}. \qquad (4)$$

The conditional state satisfies the stochastic Schrödinger equation

$$d|\bar{\bar{\psi}}_{REC}\rangle = \{-\kappa(\hat{c}^\dagger\hat{c} + \hat{A}^\dagger\hat{A})dt$$
$$+\sqrt{\kappa}[(dQ_X - idQ_Y)\hat{c} + (dQ_X + idQ_Y)\hat{A}]\}|\bar{\bar{\psi}}_{REC}\rangle, \qquad (5)$$

$|\bar{\bar{\psi}}_{REC}(0)\rangle = |\chi_{in}\rangle|\chi_{AB}^{(r)}\rangle$. Alice's measurement results are $Q_X \equiv Q_X(\infty)$ and $Q_Y \equiv Q_Y(\infty)$.

Steps 4 and 5: Mode B is displaced by $(Q_X + iQ_Y)/\sqrt{2}$. Examples of the resulting output state are shown in Figs. 2(b) and (c).

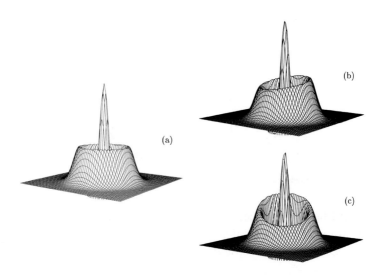

Figure 2.   CV teleportation of a two-photon Fock state: (a) Wigner function of the input state; (b) and (c) Wigner functions of the output state for two realizations of Alice's measurement results, Q_X and Q_Y. The squeezing parameter is $r = 2$.

3. Broadband Implementation

Furusawa *et al.*[5] and Bowen *et al.*[6] used an apparatus and procedure rather different from this. We model their broadband CV teleportation apparatus with the circuit sketched in Fig. 3. Significant differences are that (i) the squeezed light sources (modes a and b) emit broadband cw fields, (ii) Alice's and Victor's measurements yield photocurrents $I_X(t)$, $I_Y(t)$, and $I_{Victor}(t)$ that are continuous in time, (iii) Bob's displacement, $[I_X(t) + iI_Y(t)]/\sqrt{2}$, is executed continuously in time, and (iv) the input (mode c) is most generally a quasi-monochromatic quantum field.

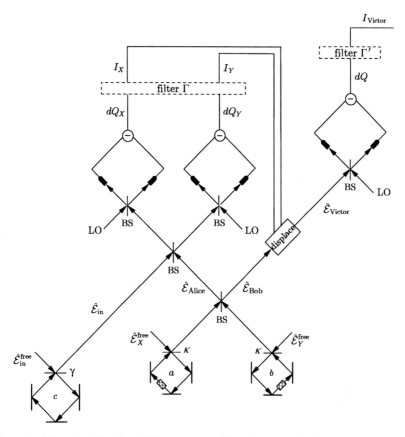

Figure 3. Schematic of broadband CV teleportation with continuous monitoring by Alice and Victor and continuous displacement by Bob; modes c, a, and b are resonant.

Within quantum trajectory theory, Alice's photocurrents are modeled by the stochastic differential equations

$$dI_X = -\Gamma(I_X dt - dQ_X), \qquad dI_Y = -\Gamma(I_Y dt - dQ_Y), \tag{6}$$

where Γ is the detection bandwidth, the bandwidth of Alice's communication to Bob, and the incremental electric charges are

$$dQ_X = 2[\sqrt{\gamma}\langle \hat{c}_X \rangle_{\text{REC}} + \sqrt{\kappa}\langle \hat{A}_X \rangle_{\text{REC}}]dt + dW_X, \tag{7a}$$

$$dQ_Y = 2[\sqrt{\gamma}\langle \hat{c}_Y \rangle_{\text{REC}} - \sqrt{\kappa}\langle \hat{A}_Y \rangle_{\text{REC}}]dt + dW_Y; \tag{7b}$$

dW_X and dW_Y are Weiner increments, as in Eqs. (2a) and (2b), and the bandwidths γ and κ of the input field and squeezing, respectively, are no longer equal to one other. As indicated in Fig. 3, the entangled fields $\hat{\mathcal{E}}_{\text{Alice}}$ and $\hat{\mathcal{E}}_{\text{Bob}}$ are produced by combining the independently squeezed fields at a 50/50 beam splitter; the respective source fields are $\sqrt{2\kappa}\hat{A}$ and $\sqrt{2\kappa}\hat{B}$, with

$$\hat{A} \equiv (\hat{a} + \hat{b})/\sqrt{2}, \qquad \hat{B} \equiv (\hat{a} - \hat{b})/\sqrt{2}. \tag{8}$$

The unnormalized conditional quantum state of the three modes c, A, and B satisfies the stochastic Schrödinger equation[7]

$$\begin{aligned}
d|\bar{\bar{\psi}}_{\text{REC}}\rangle = &\left\{\left(\hat{H}_{\text{in}}/i\hbar - \gamma\hat{c}^\dagger\hat{c}\right)dt - \kappa[\lambda(\hat{A}^\dagger\hat{B}^\dagger - \hat{A}\hat{B}) + \hat{A}^\dagger\hat{A} + \hat{B}^\dagger\hat{B}]dt \right.\\
&+(dQ_X + idQ_Y)\sqrt{\kappa}\hat{A} + (dQ_X - idQ_Y)\sqrt{\gamma}\hat{c}\\
&\left.-(I_X - iI_Y)\sqrt{\kappa}\hat{B}dt + d\hat{U}_{\text{Victor}}\right\}|\bar{\bar{\psi}}_{\text{REC}}\rangle. \tag{9}
\end{aligned}$$

The first term on the right-hand side prepares the state of input mode c, where Hamiltonian $\hat{H}_{\text{in}}$ and γ are chosen to model the quasi-monochromatic field $\hat{\mathcal{E}}_{\text{in}}$ (Fig. 3); the second term generates two-mode squeeze light, pump parameter $\lambda \leq 1$; the third and fourth arise from the backaction of Alice's measurements; the fifth from Bob's displacement; and $d\hat{U}_{\text{Victor}}$ accounts for Victor's backaction, which varies with the measurement he performs on the output field $\hat{\mathcal{E}}_{\text{Victor}}$ (Fig. 3).

4. Teleportation of the Vacuum Field

The output field is a complex object, and it is not possible to give a single measure—e.g., the usual state fidelity—of the extent to which it replicates the input. An infinite set of correlation function characterize a quasi-monochromatic quantum field. Thus, if the teleportation were perfect, any correlation function measured for $\hat{\mathcal{E}}_{\text{in}}$ would be reproduced by Victor's measurement of that correlation function for $\hat{\mathcal{E}}_{\text{Victor}}$. Of course, perfect replication is not expected: aside from the limited degree of squeezing ($\lambda < 1$), the bandwidths γ, κ, and Γ all must play a role in determining the field $\hat{\mathcal{E}}_{\text{Victor}}$. In this section we consider three of the simplest correlation functions Victor might measure and ask to what extent, according to their account, the vacuum field 'in' results in the vacuum field 'out'.

4.1. Homodyne Detection

Victor's measurement is sketched as a homodyne measurement in Fig. 3. This is the verification scheme used by Furusawa et al.[5] and Bowen et al.[6]. We model the photocurrent, I_{Victor}, by the stochastic differential equation

$$dI_{\text{Victor}} = -\Gamma'(I_{\text{Victor}}dt - dQ), \tag{10}$$

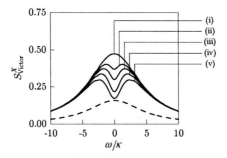

Figure 4. Broadband CV teleportation of the vacuum with homodyne detection of the output field. Victor's measured spectrum of squeezing is plotted: for $\Gamma/\kappa = 25$, $\Gamma'/\kappa = 5$, and (i) $\lambda = 0.0$, (ii) $\lambda = 0.1$, (iii) $\lambda = 0.2$, (iv) $\lambda = 0.4$, (v) $\lambda = 0.8$. The dashed curve is the shot noise spectrum for Victor's measurement of the vacuum field.

with incremental electric charge

$$dQ = 2[\sqrt{2\kappa}\langle\hat{B}_X\rangle_{\text{REC}} + I_X/\sqrt{2}]dt + dW, \tag{11}$$

where Γ' is the detection bandwidth and a third Wiener increment, dW, $\overline{dW^2} = dt$, accounts for the shot noise in Victor's measurement; we assume the X-quadrature of the field is measured, though results for the vacuum field input are quadrature independent. In the stochastic Schrödinger equation, Eq. (9), Victor's backaction is

$$d\hat{U}_{\text{Victor}} = dQ[\sqrt{2\kappa}\hat{B} + (I_X + iI_Y)/\sqrt{2}]. \tag{12}$$

Equations (6), (7a) and (7b), and (9)–(12) provide a complete quantum trajectory unraveling of broadband CV teleportation of the vacuum field, with homodyne detection of the output. They are implemented on a computer to simulate the photocurrents $I_X(t)$, $I_Y(t)$, and $I_{\text{Victor}}(t)$, and the autocorrelation

$$\overline{I_{\text{Victor}}(t)I_{\text{Victor}}(t+\tau)} = \frac{1}{N}\sum_{i=1}^{N}I_{\text{Victor}}(t_i)I_{\text{Victor}}(t_i+\tau) \tag{13}$$

is calculated, with t_i, $i = 1,\ldots,N$, a chosen set of sample times, where the number of samples is typically of order 10^6. From the correlation function an FFT gives the spectrum

$$S^X_{\text{Victor}}(\omega) = \frac{1}{2\pi}\int_{-\infty}^{\infty}d\tau\, e^{i\omega\tau}\overline{I_{\text{Victor}}(t)I_{\text{Victor}}(t+\tau)}, \tag{14}$$

which is plotted for a range of squeezing parameters in Fig. 4.

The dashed curve is the shot noise or quadrature fluctuation spectrum of the vacuum seen by Victor if he directly measures the field $\hat{\mathcal{E}}_{\text{in}}$, while curve (i) is what Victor measures for the field $\hat{\mathcal{E}}_{\text{Victor}}$ when the dispersed fields $\hat{\mathcal{E}}_{\text{Alice}}$ and

$\hat{\mathcal{E}}_{\text{Bob}}$ are not squeezed at all. Teleportation aims to recover the dashed curve for the field $\hat{\mathcal{E}}_{\text{Victor}}$. Curves (ii)–(v) show what is achieved at increasing levels of squeezing. Perfect teleportation, across the entire bandwidth of Alice's communication to Bob, can never be achieved, since the squeezing has finite bandwidth. Good recovery of the input noise can be achieved within the squeezing bandwidth, however, as curve (v), in particular, shows. Of course, Alice and Victor's detection bandwidths could be reduced to eliminate the discrepancy in the wings of the spectrum. This is a reasonable thing to do for the vacuum field input; after all, if Bob receives nothing from Alice and does nothing (no displacement) the output field is certainly the vacuum. For general quasi-monochromatic inputs, though, such bandwidth reduction exacts a price, since there is nothing akin, in the broadband scheme, to the mode-matching of Alice's measurement and the input pulse in Sec. 2. The interplay of bandwidths will become important in Sec. 5.

4.2. Heterodyne Detection

Excess noise in the wings of the spectra (Fig. 4) means that light appears at the teleporter output even though only vacuum enters the input. This light is generated by Bob's displacement. Bob's action generates a chaotic field of amplitude $[I_X(t) + iI_Y(t)]/\sqrt{2}$ across the entire bandwidth of his communication with Alice ($\Gamma/\kappa = 25$ in Fig. 4). Only at the center of the spectrum is the displacement made by Bob correlated with his received squeezed light (field $\hat{\mathcal{E}}_{\text{Bob}}$); here, but only here, a cancelation occurs to recover the vacuum via the sum in the square bracket on the right-hand side of Eq. (11).

To obtain the optical spectrum of the output light, Victor can perform a heterodyne measurement. The quantum trajectory model is very similar to the homodyne case. Victor's photocurrent is modeled by

$$dI_{\text{Victor}} = -\Gamma'(I_{\text{Victor}}dt - dQ),\tag{15}$$

with incremental electric charge

$$dQ = [\sqrt{2\kappa}\langle\hat{B}^\dagger\rangle_{\text{REC}} + (I_X - iI_Y)/\sqrt{2}]dt + dZ,\tag{16}$$

and backaction, in Eq. (9), given by

$$d\hat{U}_{\text{Victor}} = dQ[\sqrt{2\kappa}\hat{B} + (I_X + iI_Y)/\sqrt{2}].\tag{17}$$

The principal change is that the photocurrent and charge are now complex valued; the Wiener increment dZ is complex, with $\overline{dZ^2} = 0$, $\overline{dZ^*dZ} = dt$. The photocurrent autocorrelation is calculated as

$$\overline{I^*_{\text{Victor}}(t)I_{\text{Victor}}(t+\tau)} = \frac{1}{N}\sum_{i=1}^{N}I^*_{\text{Victor}}(t_i)I_{\text{Victor}}(t_i+\tau),\tag{18}$$

from which the spectrum is obtained via an FFT:

$$S_{\text{Victor}}(\omega) = \frac{1}{2\pi} \int_{-\infty}^{\infty} d\tau \, e^{i\omega\tau} \overline{I_{\text{Victor}}^*(t) I_{\text{Victor}}(t+\tau)}. \tag{19}$$

Computed spectra are plotted for a range of squeezing parameters in Fig. 5(a). The dashed line represents the shot noise in heterodyne detection. For the optical spectra we subtract this shot noise from the computed results—curves (i)–(v)— to obtain the spectra displayed in Fig. 5(b). In curve (v), in particular, a good approximation to the vacuum is recovered for frequencies near $\omega/\kappa = 0$. More generally, away from the center of the squeezing spectrum, real photons appear at the teleporter output. The full bandwidth of this emitted light is $\Gamma/\kappa = 25$, the chosen bandwidth of Alice's communication with Bob. The widths of the spectra in the figure are limited by Victor's detection bandwidth $\Gamma'/\kappa = 5$.

4.3. Photoelectron Counting

Our third example of a possible measurement by Victor is photoelectron counting; specifically, we consider a measurement of the intensity correlation function

$$g_{\text{Victor}}^{(2)}(\tau) \equiv \frac{\langle \hat{\mathcal{E}}_{\text{out}}^\dagger(0) \hat{\mathcal{E}}_{\text{out}}^\dagger(\tau) \hat{\mathcal{E}}_{\text{out}}(\tau) \hat{\mathcal{E}}_{\text{out}}(0) \rangle}{\langle \hat{\mathcal{E}}_{\text{out}}^\dagger(0) \hat{\mathcal{E}}_{\text{out}}(0) \rangle^2}, \tag{20}$$

where for a stationary field the origin of time is unimportant and set to zero. It is necessary that Victor only detect frequencies near the center of the squeezing spectrum, otherwise his measurement will be contaminated by the excess light generated by Bob's displacement. Thus, he detects the field $\hat{\mathcal{E}}_{\text{out}}$, source field $\sqrt{\Gamma'}\hat{d}$, at the output of the filter cavity depicted in Fig. 6. To replace Eqs. (10)

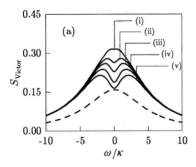
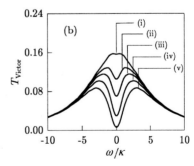

Figure 5. Broadband CV teleportation of the vacuum with heterodyne detection of the output field. Victor's measured heterodyne current spectrum (a) and optical spectrum (b) are plotted: for $\Gamma/\kappa = 25$, $\Gamma'/\kappa = 5$, and (i) $\lambda = 0.0$, (ii) $\lambda = 0.1$, (iii) $\lambda = 0.2$, (iv) $\lambda = 0.4$, (v) $\lambda = 0.8$. The dashed curve is the shot noise spectrum.

and (15), which model the measured photocurrents in homodyne and heterodyne detection, the quantum trajectory unraveling for photoelectron counting includes two kinds of quantum jump:

$$|\bar{\bar{\Psi}}_{REC}\rangle \rightarrow \sqrt{\Gamma'}\hat{d}|\bar{\bar{\Psi}}_{REC}\rangle, \qquad (21a)$$

$$|\bar{\bar{\Psi}}_{REC}\rangle \rightarrow [\sqrt{\Gamma'}\hat{a} + \sqrt{2\kappa}\hat{B} + (I_X + iI_Y)/\sqrt{2}]|\bar{\bar{\Psi}}_{REC}\rangle, \qquad (21b)$$

with the first (second) occuring for a photoelectron count recorded in the transmitted (reflected) channel of the filter cavity (Fig. 6). Between jumps, the stochastic Schrödinger equation, Eq. (9), is solved, with Victor's backaction

$$d\hat{U}_{Victor} = -\{\Gamma'\hat{d}^\dagger\hat{d} + \sqrt{\Gamma'}\hat{d}^\dagger[\sqrt{2\kappa}\hat{B} + (I_X + iI_Y)/\sqrt{2}]\}dt. \qquad (22)$$

Equations (6), (7a) and (7b), (9), and (21a)–(22) provide a complete quantum trajectory unraveling of broadband CV teleportation of the vacuum field with photoelectron counting at the output. We have used them to simulate a single ongoing trajectory and to compute the intensity correlation function, Eq. (20), as

$$g^{(2)}_{Victor}(\tau) = \langle\hat{d}^\dagger\hat{d}\rangle^{-1}\frac{1}{N}\sum_{i=1}^{N}\langle(\hat{d}^\dagger\hat{d})(t_i + \tau)\rangle_{REC}, \qquad (23)$$

where t_i, $i = 1,\ldots,N$, denote the times of quantum jumps in transmission [Eq. (21a)] and $\langle\hat{d}^\dagger\hat{d}\rangle$ is the time average of the photon number expectation $\langle(\hat{d}^\dagger\hat{d})(t)\rangle_{REC}$.

Results obtained for a range of different squeezing parameters are plotted in Fig. 7. Perhaps more important than $g^{(2)}(\tau)$ itself are the mean photon numbers in Victor's filter cavity shown to the right in the figure. These numbers provide

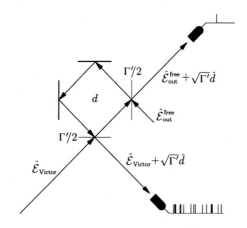

Figure 6.   Filter cavity for Victor's photoelectron counting.

a measure of the contamination which the teleportation of any non-vacuum input field will face. Perfect teleportation of the vacuum across the filter bandwidth $(\Gamma'/\kappa = 0.2)$ would yield $\langle \hat{d}^\dagger \hat{d} \rangle = 0$. Without squeezing, there is on average, one half a photon in the filter cavity; as already noted, Alice's shot noise (her measurement of the input vacuum fluctuations) is realized at the teleporter output, through the mediation of Bob's displacement, as real light. With increasing squeezing, the photon number decreases, approaching the ideal $\langle \hat{d}^\dagger \hat{d} \rangle = 0$. There is also a decrease in the correlation time of the intensity fluctuations, which, to a good approximation at least, emulate those of chaotic light, i.e., with $g^{(2)}_{\text{Victor}}(0) = 2$.

5. Teleportation of Resonance Fluorescence

We turn now to the teleportation of a non-vacuum input field. Single-atom resonance fluorescence is chosen as the ideal example. It presents a number of advantages: (i) its spectrum is easily recognized, undergoing the familiar development from Rayleigh scattering to the Mollow triplet as the Rabi frequency is increased;[8] (ii) its photon counting sequences are nonclassical, i.e., antibunched;[9,10] and (iii) computational requirements are kept at a minimum, as the size of the Hilbert space increases by a factor of two only over the vacuum field case. We seek to teleport the Mollow spectrum (heterodyne detection by Victor, Sec. 4.2) and photon antibunching (photoelectron counting, Sec. 4.3). The stochastic Schrödinger equation, Eq. (9), now has input mode Hamiltonian

$$\hat{H}_{\text{in}} = i\hbar(\Omega/2)(\hat{\sigma}_+ - \hat{\sigma}_-), \qquad (24)$$

where Ω is the Rabi frequency. We assume the fluorescence is coupled into Alice's detectors with unit efficiency; thus, the parameter γ in Eqs. (7a) and (7b) is the

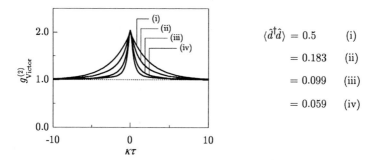

Figure 7.   Broadband CV teleportation of the vacuum with photoelectron counting of the output field. Victor's measured intensity correlation function is plotted: for $\Gamma/\kappa = 10$, $\Gamma'/\kappa = 0.2$, and (i) $\lambda = 0.0$, (ii) $\lambda = 0.3$, (iii) $\lambda = 0.5$, (iv) $\lambda = 0.7$. The mean photon number in Victor's filter cavity is shown to the right.

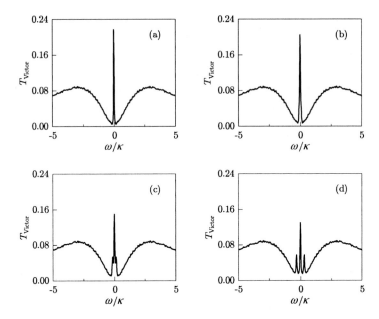

Figure 8.   Broadband CV teleportation of the Mollow spectrum. Resonance fluorescence is injected at the teleporter input and Victor's measured optical spectrum is computed: for $\Gamma/\kappa = 25$, $\Gamma'/\kappa = 5$, $\lambda = 0.8$, atomic halfwidth $\gamma/\kappa = 0.025$, and Rabi frequency (a) $\Omega/\kappa = 0.04$, (b) $\Omega/\kappa = 0.08$, (c) $\Omega/\kappa = 0.16$, (d) $\Omega/\kappa = 0.32$.

atomic linewidth (halfwidth), as in Eq. (9).

5.1. *Mollow Spectrum*

Examples of teleported spectra are shown in Fig. 8. The atomic linewidth, $\gamma/\kappa = 0.025$, is chosen so that the fluorescence spectrum fits comfortably within the squeezing bandwidth—i.e., within the central dip of the shot noise spectrum communicated by Alice to Bob [Fig. 5(b) curve (v)]. The development from Rayleigh scattering to the Mollow triplet is well reproduced in Fig. 8. There is excess light, contamination, in the wings of the spectra though, due to the finite bandwidth of the squeezing and large bandwidth chosen for Alice's communication ($\Gamma/\kappa = 25$). While the contamination can be reduced, by reducing Alice's bandwidth, it cannot be eliminated altogether. In this regard, it is important to note that the teleportation attempted here differs in a fundamental way from the elementary scheme of Sec. 2; the input field, squeezed (entangled) light, and Alice's detection and communication with Bob are not all mode matched. Thus, while reducing Alice's bandwidth reduces the contamination in the wings of the

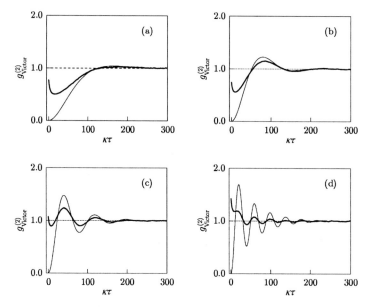

Figure 9. Broadband CV teleportation of photon antibunching. Resonance fluorescence is injected at the teleporter input and Victor's measured intensity correlation function is computed (bold line): for $\Gamma/\kappa = 2$, $\Gamma'/\kappa = 0.05$, $\lambda = 0.7$, atomic halfwidth $\gamma/\kappa = 0.0125$, and Rabi frequency (a) $\Omega/\kappa = 0.02$, (b) $\Omega/\kappa = 0.04$, (c) $\Omega/\kappa = 0.08$, (d) $\Omega/\kappa = 0.16$. The light line is the correlation function at the input.

teleported spectrum, it eventually fails to transmit the correlations needed if Bob's displacement is to maintain the squeezing dip where the fluorescence spectrum sits; eventually (e.g. for $\Gamma/\kappa = 1$) Victor's measured spectrum becomes highly distorted over the range of the Mollow peaks. Such bandwidth effects are particularly important when teleportation of photon antibunching is considered.

5.2. *Photon Antibunching*

Figure 9 presents results for the intensity correlation function at the output of Victor's filter cavity (Fig. 6) computed using the procedure outlined in Sec. 4.3. In order to minimize contamination by excess light in the spectral wings, Alice and Victor's bandwidths are significantly reduced and the atomic linewidth is set to half its value in Fig. 8. The results presented in the figure are mixed. At weak excitation, in frames (a) and (b), photon antibunching $[g^{(2)}_{\text{Victor}}(0) < 1]$ is preserved in the output field. It is lost at higher excitation, however; for example, although the correlation function in frame (c) is nonclassical by a different measure— $g^{(2)}_{\text{Victor}}(\tau) > g^{(2)}_{\text{Victor}}(0)$ is nonclassical—photon antibunching is no longer present.

At still higher excitation, frame (d), the correlation function is classical.

The generally poor fidelity is caused by the bandwidth constraints, combined with the imperfect teleportation of the vacuum field demonstrated by the photon numbers to the right in Fig. 7. The bandwidth of resonance fluorescence increases with increasing Rabi frequency; hence, the finite bandwidth of the squeezing is more detrimental in Fig. 9(d) than in Fig. 9(a). We might suppose that reducing the atomic linewidth would alleviate this problem; but a reduced linewidth means lower input photon flux and a smaller photon number in Victor's filter cavity. Leaving all else unchanged, the background light level is the same. The background light (which produces the "hook" at small τ in Fig. 9) becomes more detrimental.

Clearly severe compromises must be made in order to optimize the results. This is difficult to do numerically. Our aim in future work is to derive an analytical expression for $g^{(2)}_{\text{Victor}}(\tau)$ which will facilitate a full study of bandwidth constraints and their interplay with the level of squeezing.

6. Conclusions

We have developed a quantum trajectory formalism for broadband continuous variable (CV) teleportation. The formalism is multi-mode, with inputs and outputs viewed as quasi-monochromatic quantum fields rather than single-mode quantum states. The multi-mode aspect goes beyond the proposal for CV teleportation of Braunstein and Kimble,[4] but is a notable feature of the proposal's experimental implementation.[5,6] The implementation is broadband with no temporal mode matching between the input field, squeezed light, and homodyne measurements of Alice; mode matching to the input is replaced by mode selection (filtering) at the output.

The need to filter excess light from the teleporter output places severe constraints on the teleportation of quantum fields. We illustrated the issues involved with the example of single-atom resonance fluorescence, for which teleportation of the Mollow spectrum and photon antibunching was demonstrated. In the latter case, the transfer of photon antibunching from input to output was demonstrated for weak excitation only, and the fidelity of the transfer was rather poor. A thorough exploration of limitations on the teleportation of quantum fields is planned for future work.

Acknowledgments

This work was supported by the Marsden Fund of the RSNZ.

References

1. C. H. Bennett, G. Brassard, C. Crépeau, R. Jozsa, A. Peres, and W. K. Wooters, *Phys. Rev. Lett.* **93**, 1895 (1993).
2. D. Bouwmeester, J.-W. Pan, K. Mattle, M. Eibl, H. Weinfurter, and A. Zeilinger, *Nature* **390**, 575 (1997).
3. L. Vaidman, *Phys. Rev. A* **49**, 1473 (1994).
4. S. L. Braunstein and H. J. Kimble, *Phys. Rev. Lett.* **80**, 869 (1998).
5. A. Furusawa, J. L. Sørensen, S. L. Braunstein, C. A. Fuchs, H. J. Kimble, and E. S. Polzik, *Science* **282**, 706 (1998).
6. W. P. Bowen, N. Treps, B. C. Buchler, R. Schnabel, T. C. Ralph, H.-A. Bachor, T. Symul, and P. K. Lam, *Phys. Rev. A* **67**, 032302 (2003).
7. H. J. Carmichael, "Quantum trajectory theory of continuous variable teleportation," in *Fluctuations and Noise in Photonics and Quantum Optics II*, eds. P. Heszler, D. Abbott, J. R. Gea-Banacloche, and P. R. Hemmer, Proceedings of SPIE, Vol. 5468, (SPIE, Bellingham, WA, 2004) pp. 282-91.
8. B. R. Mollow, Phys. Rev. **188**, 1969 (1969).
9. H. J. Carmichael and D. F. Walls, J. Phys. B **9**, L43, 1199 (1976).
10. H. J. Kimble, M. Dagenais, and L. Mandel, Phys. Rev. Lett. **39**, 691 (1977).

SILICON-BASED NUCLEAR SPIN QUANTUM COMPUTER

HSI-SHENG GOAN

Department of Physics, National Taiwan University, Taipei 106 Taiwan (ROC)

We review the basic physics and operation principles of the silicon-based quantum computer proposed by Kane, one of the most promising solid-state quantum computer proposals. We describe in some details how single- and two-qubit operations and readout measurements can, in principle, be performed for the Kane quantum computer. In addition, we also mention briefly its recent theoretical progress and development.

1. Introduction

Advances in quantum algorithms which outperform their best known classical counterparts and quantum error correction codes which protect quantum information against noise have prompted a search for a practical quantum computer recently. Many different solid-state (mesoscopic) quantum computer architectures, such as superconducting Josephsen junction devices, impurities in semiconductors, quantum dots, have been proposed. Among them, the silicon-based quantum computers have received much attention due to their potential advantage to exploit the existing strength of silicon technology in semiconductor industry and their considerable ability to scale up to large scale.

In this paper, we review in some details the basic physics and operation principles of the silicon-based quantum computer proposed by Kane [1,2], one of the most promising solid-state quantum computer proposals. In the Kane's proposal, the nuclear spins of impurity ^{31}P donors placed in a regular array in pure silicon crystal are used as the quantum bits (qubits). The interaction between qubits of donor nuclear spins are mediated through the donor electrons. The qubit manipulations and quantum gate operations are controlled by external magnetic field, surface gate voltage, and nuclear magnetic resonance (NMR) pulses. In this paper we also mention briefly the recent theoretical development on this silicon-based electron-mediated nuclear spin quantum computer. Its recent experimental progress has been reported in the review article of Ref. [3].

2. Shallow donor in silicon and operation condition

One of the considerations for choosing nuclear spins of ^{31}P in silicon as qubits is that $I = 1/2$ phosphorus nuclear spins are extremely well isolated from their environment [1]. Their silicon host can in principle be purified to contain only $I = 0$ stable isotopes. As a result, the relaxation and decoherence times of donor nuclear and electron spins are comparatively long. For example, the nuclear spin relaxation times were measured to be $T_{1n} > 10$ hours at a temperature of $T = 1.25$K, $B = 3.2$T, and the electron relaxation time was measured about $T_{1e} \approx 30$ hours under similar conditions. The nuclear spin decoherence time T_{2n} were reported to be larger than or in the order of few seconds and the electron decoherence time to be $T_{2_e} = 62$ms at $T = 6.9K$. At milli-Kelvin temperatures and for a specifically engineered donor system, the decoherence time is likely to be even longer. The electron decoherence time is much larger than the typical time scale given by the inverse of electron-electron exchange interaction in P:Si system (see Table 1). The nuclear spin (qubit) decoherence times is much longer than the gate operation times, typically tens of μs for the Kane quantum computer. These relatively long decoherece times fulfill an important requirement for the Kane proposal to be considered as a practical quantum computer architecture.

The replacement of a silicon atom by a donor ^{31}P in a silicon host can occur easily since the atoms have approximately the same size. The phosphorus atom has five valence electrons. We can to first approximation assume that four of these electrons, which fill states rather similar to those of silicon, will participate in four covalent bonds with the four neighbor atoms. The extra remaining electron, at low temperature, is normally bound to the phosphorus atom which has an additional nuclear charge $+e$. Thus a P atom behaves effectively like a hydrogen-like atom embedded in silicon. One may obtain an order of magnitude estimate for the donor Bohr radius, a_B^*, and bound state energies, E_n, using the following hydrogen-like atom formula: $a_B^* = \varepsilon a_B(m_e/m^*)$, and $E_n = (E_n^H/\varepsilon^2)(m_e/m^*)$. Given that the static dielectric constant and effective electron mass in Si: $\varepsilon = 11.7$ and $m^* \approx m_T^* = 0.2m_e$ (where m_e is the free electron mass, and m_T^* is the transverse effective mass described later), as well as the Bohr radius and bound state energies for hydrogen atom: $a_B = 0.53$Å, and $E_n^H = -13.6$eV$/n^2$ with n an positive integer, we obtain $a_B^* \approx 30$Å and $E_1 \approx -20$meV. The estimated values of E_n indicate that the P donor occupies a energy level at a distance below the conduction band minimum, small compared with the (indirect) conduction-valence energy band gap ($E_g = 1.12$eV in Si); it is thus called a shallow donor. One of the purposes of the electrons in Kane's proposed quantum computer is to mediate nuclear spin coupling. At sufficient low temperatures, electrons only occupy the lowest energy bound states ($1s$ orbitals or band) at the donors. The $1s$ electron wave function is concentrated at the donor nucleus, yielding a large hyperfine interaction energy. The estimated values of a_B^* implies that the donor electron wave function extends tens or hundreds of angstroms away from the donor nucleus, allowing electron-mediated nuclear spin coupling to occur over comparable distances.

There exists, however, more complexity in the real situation for silicon semiconductor [4]. The lowest conduction band minimum in Si is in the direction [100], and by (cubic) symmetry in other equivalent [100] directions. There are thus six conduction band minima near the zone boundaries around $|k| \sim 0.85(2\pi/a)$, where a is the lattice constant of Si. The prolate ellipsoid of constant energy near each conduction band valley has two equal transverse axes. The dispersion relation, for example, has the form

$$E(\mathbf{k}) = \frac{\hbar^2}{2} \left(\frac{k_x^2 + k_y^2}{m_T^*} + \frac{(k_z - k_{z0})^2}{m_L^*} \right) \tag{1}$$

for the ellipsoids [001] and [00$\bar{1}$], where $k_{z0} = 0.85(2\pi/a)$. There appears a longitudinal effective mass $m_L^* = 0.98\,m_e$ and transverse mass $m_T^* = 0.2\,m_e$. To obtain more precise bound state energy levels for ^{31}P donor electron in Si, one should take into account the anisotropy of the conduction band, the so-called central cell correction and the interaction between the six degenerate valleys, known as the valley-orbit coupling [4]. As a result, the $1s$ degenerate ground states of the six equivalent [100] valleys in Si splits into a singlet of A_1 symmetry, a triplet of T_2 symmetry and a doublet of E symmetry. The shallow donor singlet ground state [4] has an energy about -45.5 meV below the Si conduction band edge and the lowest excited state, the triplet, is approximately 13 meV above the singlet ground state. This provide a condition to ignore high-lying single-electron state of the donor atom if the temperature T is such that $k_B T \ll \Delta E = 13$ meV, where k_B is the Boltzmann constant. This condition is well satisfied, as the operation temperature, described in next section, for the proposed quantum computer is roughly at $T = 100$ mK ($k_B T = 0.0086$ meV). Thus we may treat the effective low-energy and low-temperature Hamiltonian involving only the spin degrees of freedom of the system.

We now describe the basic operation conditions for the Si:^{31}P quantum computer system. Throughout the computation the electrons must be in a non-degenerate ground state to avoid irreversible interactions between electron and nuclear spins occurring as the computation proceeds. An external magnet field, **B**, is applied to break the shallow donor electron ground state two-fold spin degeneracy. At sufficient low temperature T, the electron will only occupy the lowest energy spin level when the electron Zeeman splitting is much larger than the thermal energy, $g_e \mu_B B \gg k_B T$, where μ_B is the Bohr magneton, and $g_e \approx 2$ is the Lande g-factor in Si. At $T = 100$mK and $B = 2$ tesla $(2\mu_B B = 0.23$meV$)$, the electrons will be completely spin-polarized: $(n_\uparrow^e / n_\downarrow^e) \approx \exp[-2\mu_B B/(k_B T)] \approx 2.14 \times 10^{-12}$, where n^e represents the donor electron number density. These conditions however do not fully polarize the nuclear spins: $(n_\downarrow^n / n_\uparrow^n) \approx \exp[-2g_n \mu_n B/(k_B T)] \approx 0.98$, where n^n stands for the donor nuclear number density, the nuclear g-factor $g_n = 1.13$ for ^{31}P, μ_n is the nuclear magneton and $2g_n \mu_n B \approx 0.00014$ meV. The polarizations of the nuclear spins are instead determined by interactions with the polarized electrons.

In Table 1, we list some relevant and typical energy scales in mini electron Volts (meV) and their corresponding frequencies in Hertzes (Hz) for the silicon-based electron-mediated nuclear spin quantum computer.

Table 1.   Relevant energy scales for the silicon-based electron-mediated nuclear spin quantum computer proposed by Kane. Here * represents that these energies can be controlled externally by varying the A-gate and J-gate voltages.

energy	meV	Hz
indirect conduction-valence energy band gap E_g of Si	1120	2.7×10^{14}
electron ground state energy of ^{31}P donor in Si	-45.5	-1.1×10^{13}
electron 1st excited state energy of ^{31}P donor in Si	-33.9	-8.2×10^{12}
2nd electron binding energy in D^- state of ^{31}P donor in Si	-1.7	-4.1×10^{11}
temperature energy scale $k_B T$ at $T = 100$mK	0.0086	2.1×10^9
electron Zeeman energy $\mu_B B$ at $B = 2$T	0.116	2.8×10^{10}
nuclear Zeeman energy $g_n \mu_n B$ at $B = 2$T	7.1×10^{-5}	1.7×10^7
*typical hyperfine interaction A	1.2×10^{-4}	2.9×10^7
*typical nuclear resonance energy $h\nu_A$ for single qubit	3.8×10^{-4}	9.3×10^7
nuclear full width at half maxima $4g_n \mu_n B_{ac}$ at $B_{ac} = 10^{-3}$T	1.4×10^{-7}	3.4×10^4
*typical electron exchange energy $4J$	0.124	3.0×10^{10}
*typical nuclear exchange energy $h\nu_J$	3.1×10^{-7}	7.5×10^4

3. Single-qubit operations

The effective low-energy and low-temperature Hamiltonian for a ^{31}P nuclear spin-electron system in Si under a uniform magnetic field $B \parallel z$ can be written as $H_{en} = \mu_B B \sigma_z^e - g_n \mu_n B \sigma_z^n + A \sigma^e \cdot \sigma^n$, where e and n appearing in the superscripts and subscripts represent quantities for the electron and the nucleus respectively, σ's are Pauli matrices, and $A = 8\pi \mu_B g_n \mu_n |\Psi(0)|^2/3$ is the contact hyperfine interaction energy with $|\Psi(0)|^2$ the probability density of the electron wave function evaluated at the position of the nucleus. Direct diagonalization of this simple Hamiltonian gives the eigen energies, with each associated eigen state as a subscript (e.g., $E_{|en\rangle}$ stands for the energy of the state $|en\rangle$), as follows: $E_{|\uparrow 0\rangle} = \mu_B B - g_n \mu_n B + A$, $E_{\alpha|\uparrow 1\rangle + \gamma|\downarrow 0\rangle} = \sqrt{(\mu_B B + g_n \mu_n B)^2 + (2A)^2} - A$, $E_{-\gamma|\uparrow 1\rangle + \alpha|\downarrow 0\rangle} = -\sqrt{(\mu_B B + g_n \mu_n B)^2 + (2A)^2} - A$, and $E_{|\downarrow 1\rangle} = -\mu_B B + g_n \mu_n B + A$. Here the eigen states are written in electron σ_z^e and nuclear σ_z^n basis. For example, $|en\rangle = |\downarrow\, 0\rangle$ represents the electron spin down ($\downarrow$) and nuclear spin up (0) state. The coefficients α and γ are: $\alpha = [1 + (\Lambda - \sqrt{1 + \Lambda^2})^2]^{-1/2}$, and $\gamma = [1 + (\Lambda + \sqrt{1 + \Lambda^2})^2]^{-1/2}$, where $\Lambda = (\mu_B B + g_n \mu_n B)/(2A)$. We can see that if Λ is very large, which is usually the case for the proposed quantum computer, then $\alpha \to 1$ and $\gamma \to 0$.

For the sake of the more complicated two-qubit Hamiltonian encountered later, let us nevertheless calculate the energy levels perturbatively, by treating the hyperfine interaction, $H' = A\sigma^e \cdot \sigma^n$, as a perturbation to obtain: $E_{|\uparrow 0\rangle}^{(2)} = \mu_B B - g_n \mu_n B + A$, $E_{|\uparrow 1\rangle}^{(2)} = \mu_B B + g_n \mu_n B - A + 2A^2/(\mu_B B + g_n \mu_n B)$, $E_{|\downarrow 1\rangle}^{(2)} = -\mu_B B + g_n \mu_n B + A$, and $E_{|\downarrow 0\rangle}^{(2)} = -\mu_B B - g_n \mu_n B - A - 2A^2/(\mu_B B + g_n \mu_n B)$. If the electron is in its spindown polarized ground state, the frequency separation between the two associated energy levels, $|\downarrow 1\rangle$ and $|\downarrow 0\rangle$, is found to be

$$h\nu_A = 2g_n \mu_n B + 2A + 2A^2/(\mu_B B + g_n \mu_n B), \tag{2}$$

The above perturbative results can also be obtained to the second order in A by directly expanding the exact eigen energies, up to the second power of Λ^{-1}. In addition, the coefficients of the states, also by direct expansion of α and γ to the first power of Λ^{-1} and by using the relation $\alpha^2 + \gamma^2 = 1$, have the approximate values: $\alpha = \{1 - [A/(\mu_B B + g_n \mu_n B)]^2\}^{1/2}$, and $\gamma = A/(\mu_B B + g_n \mu_n B)$. This result can be obtained by calculating the first-order wave function shift using the perturbation theory. In a Si:^{31}P system, $2A/h = 58$MHz, and $A > g_n \mu_n B$ for $B < 3.5$T.

Applying the A-gate voltage (on the top of the donor atom, see Fig. 3) shifts the electron wave function envelope away from the nucleus and reduces the hyperfine interaction. This hyperfine interaction shift and the corresponding nuclear resonance frequency change, Eq. (2), as a function of applied voltage, is estimated and shown in Fig. 2 in Kane's original paper [1]. A more detailed and refined calculations that take into account simultaneously the anisotropy of the effective masses in the silicon host, the electric field potential and the interface regions in the Si wafer device can be found in Refs. [5].

Since the hyperfine interaction and thus the nuclear resonance frequency is controllable externally, the nuclear spins can be selectively brought into resonance with a globally applied a.c. magnetic field, B_{ac}, allowing arbitrary rotations along the x and y axes to be performed on each nuclear spin. This is similar to the nuclear magnetic resonance (NMR) technique but with extra selectivity to address individual qubits. If the separation frequency (Larmor frequency) between $|\downarrow 1\rangle$

and $|\downarrow 0\rangle$ states of a targeted donor atom is tuned, with $A = A_x$ in Eq. (2), to the resonance frequency of the externally applied a.c. magnetic field, and the temperature is low enough, we may, to an approximation, consider only the polarized electron ground state subspace. In this case, to the first order, the speed of the rotation frequency (Rabi frequency) along the x axis may be approximated by $h\nu_r = g_n\mu_n B_{ac}[1 + A/(g_n\mu_n B)]$.

The rotation along the z axis can be achieved by tuning the hyperfine interaction of a targeted donor atom from A to A_z and allowing free evolution, but without applying an a.c. magnetic field [10]. This gives a difference in the Larmor precession frequency around the z axis with respect to other unperturbed donor atoms: $h\nu_z = 2(A - A_z) + 2(A^2 - A_z^2)/(\mu_B B + g_n\mu_n B)$. The speed of single-qubit z-axis rotation depends on how much it is possible to vary the strength of the hyperfine interaction A. Any single-qubit gate can thus be performed by means of a product of the rotations around x, y and z axes.

4. Two-qubit system

The effective Hamiltonian for two coupled donor nuclear spin-electron systems, valid at energy scales small compared to the donor-electron binding energy, is

$$H_{coup} = \mu_B B \sigma_z^{1e} - g_n\mu_n B \sigma_z^{1n} + \mu_B B \sigma_z^{2e} - g_n\mu_n B \sigma_z^{2n}$$
$$+ A_1 \sigma^{1e} \cdot \sigma^{1n} + A_2 \sigma^{2e} \cdot \sigma^{2n} + J \sigma^{1e} \cdot \sigma^{2e}, \qquad (3)$$

where A_1 and A_2 are the hyperfine interaction energies of the respective nucleus-electron systems and $4J$, the exchange energy, depends on the overlap of the electron wave functions. In order for exchange coupling between the electron spins to be significant, the separation between donors should not be too large. As shall be seen below, significant coupling between nuclei will occur when $4J \approx 2\mu_B B$. This condition sets an approximately necessary separation between donors. To get an estimate about the necessary separation, let us use the formula for exchange interaction, derived for well separated H-H atoms but with values appropriate for ^{31}P donors in Si: $4J(r) \approx 1.6[e^2/(\varepsilon a_B^*)][r/a_B^*]^{5/2}\exp(-2r/a_B^*)$, where r is the distance between donors. One can compute that when $4J \approx 2\mu_B B = 56$GHZ, the separation between donors is roughly located in $100 - 200$Å. Since the value of J depends on the electron wave function overlap, it can be varied by an J-gate positioned between the donors (see Fig. 3). In practice, the exchange interaction is complicated in Si because the contribution from each valley interferes, leading to oscillatory behavior [6,7] of $J(r)$.

In a magnetic field, the $|\downarrow\downarrow\rangle$ state will be the electron ground state if $k_B T \ll J < \mu_B B/2$. In the polarized $|\downarrow\downarrow\rangle$ electron states, the energies of the nuclear states can be calculated to the second order in A using the perturbation theory. We diagonalize the unperturbed Hamiltonian and then find the non-zero matrix elements of the perturbation,

$$H' = A_1 \sigma^{1e} \cdot \sigma^{1n} + A_2 \sigma^{2e} \cdot \sigma^{2n}, \tag{4}$$

with respect to the unperturbed eigen states. Let us introduce the following notations for the electron and nuclear states for abbreviation: $|s_e\rangle = |\uparrow\downarrow + \downarrow\uparrow\rangle/\sqrt{2}$, $|a_e\rangle = |\uparrow\downarrow - \downarrow\uparrow\rangle/\sqrt{2}$, $|s_n\rangle = |10 + 01\rangle/\sqrt{2}$, $|a_n\rangle = |10 - 01\rangle/\sqrt{2}$. We focus on the nuclear spin energy levels in the spin-polarized electron ground states. For simplicity, let us consider the case where $A_1 = A_2 = A$ in H', Eq. (4). Using the second-order energy shift formula, we find that the electron-nuclear spin energy levels in the electron spin-polarized $|\downarrow\downarrow\rangle$ states, to the second order in A, are:

$$E^{(2)}_{|\downarrow\downarrow\rangle|11\rangle} = -2\mu_B B + 2g_n\mu_n B + J + 2A, \tag{5}$$

$$E^{(2)}_{|\downarrow\downarrow\rangle|s_n\rangle} = -2\mu_B B + J - \frac{2A^2}{\mu_B B + g_n\mu_n B}, \tag{6}$$

$$E^{(2)}_{|\downarrow\downarrow\rangle|a_n\rangle} = -2\mu_B B + J - \frac{2A^2}{\mu_B B + g_n\mu_n B - 2J}, \tag{7}$$

$$E^{(2)}_{|\downarrow\downarrow\rangle|00\rangle} = -2\mu_B B - 2g_n\mu_n B + J - 2A$$
$$- \frac{2A^2}{\mu_B B + g_n\mu_n B} - \frac{2A^2}{\mu_B B + g_n\mu_n B - 2J}. \tag{8}$$

The $|\downarrow\downarrow\rangle|a_n\rangle$ state is lowered in energy with respect to $|\downarrow\downarrow\rangle|s_n\rangle$ by:

$$h\nu_J = 2A^2 \left(\frac{1}{\mu_B B + g_n\mu_n B - 2J} - \frac{1}{\mu_B B + g_n\mu_n B} \right). \tag{9}$$

The $|\downarrow\downarrow\rangle|11\rangle$ state is above the $|\downarrow\downarrow\rangle|s_n\rangle$ state and the $|\downarrow\downarrow\rangle|00\rangle$ state below the $|\downarrow\downarrow\rangle|a_n\rangle$ state by an energy $h\nu_A$, Eq. (2). For the Si:^{31}P system at $B = 2$T and for $4J/h = 30$GHz, Eq. (9) yields the effective nuclear exchange frequency $\nu_J \approx 75$ KHz. This nuclear spin exchange frequency approximates the rate at which binary operations can be performed on the computer.

5. Controlled-NOT operation

The controlled-NOT operation (conditional rotation of the target spin by 180°) can be realized [2,8] by the combined applications of B_{ac} and adiabatic variations in J and $\Delta A = A_1 - A_2$, in which the gate biases are varied slowly. The sequence of the adiabatic steps performed and the associated evolution of nuclear spin states and energy levels for the controlled-NOT operation are schematically illustrated in Fig. 1.

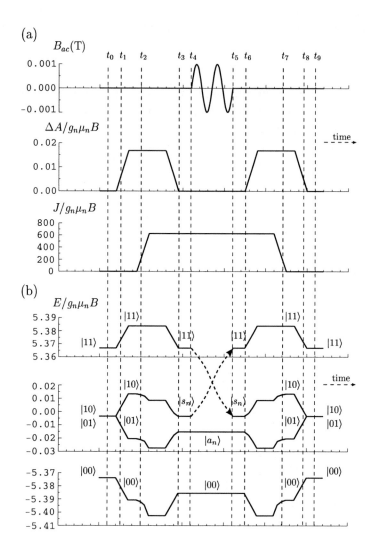

Figure 1. The controlled-NOT operation realized by the combined applications of B_{ac} and adiabatic variations in J and $\Delta A = A_1 - A_2$. (a) The sequence of controls are illustrated schematically. All energy scales are in units of nuclear Zeeman energy: $g_n \mu_n B$. (b) The corresponding evolution of the nuclear spin states and energy levels are presented. Throughout the controlled-NOT operation $J < \mu_B B/2$. As a consequence, the electron spin state of the system is always in polarized $|\downarrow\downarrow\rangle$ state and only the associated nuclear spin states are explicitly shown. Note that the energy levels are presented at the same scale but in different energy intervals, and they are obtained by keeping $A_2 \approx 1.683\, g_n \mu_n B \approx 0.001\, \mu_B B$ fixed and varying $A_1 = A_2 + \Delta A$.

At $t = t_0$, indicated in Fig. 1 , the two spin systems are decoupled ($J=0$) and $\Delta A = 0$ so that $|10\rangle$ and $|01\rangle$ are degenerate. At t_1, $\Delta A > 0$ ($A_1 > A_2$) breaks this degeneracy ($|10\rangle$ above $|01\rangle$) and distinguishes the control qubit from the target qubit. At t_2, J is turned on and in the case when $\Delta A \gg h\nu_J$, the eigenstates do not evolve away from themselves much, i.e., they roughly remain in the same states as they were. When ΔA is slowly decreased to zero with J on, the $|10\rangle$ state evolves adiabatically into $|s_n\rangle$ state and $|01\rangle$ into $|a_n\rangle$. The energy splitting between the central levels at t_3 then reduce to the effective nuclear spin exchange frequency $h\nu_J$ between $|s_n\rangle$ and $|a_n\rangle$ states. At t_4, a linear polarized magnetic field B_{ac} is applied, for example, in the x-direction resonant with the $|11\rangle$ — $|s_n\rangle$ gap. At first sight, B_{ac} seems to be also resonant with the $|00\rangle$ — $|a_n\rangle$ gap; however, to the first order in time-dependent perturbation theory, the matrix element of this transition is zero since the nuclear singlet state is not coupled to the other triplet states by the perturbation: $g_n\mu_n(\sigma_x^{1n} + \sigma_x^{2n})B_{ac}\cos(\omega t)$. B_{ac} is left on until t_5, when it has transformed $|11\rangle$ into $|s_n\rangle$ and vice versa. The $|s_n\rangle$ and $|a_n\rangle$ states are then adiabatically transformed back into $|10\rangle$ and $|01\rangle$ in a reverse of the sequence of steps performed at the beginning of the operation. We can see that the qubits whose energy levels vary during the adiabatic procedure are unchanged. However, the states are inverted if and only if the control qubit (first qubit) is $|1\rangle$. Therefore the controlled-NOT operation has been performed. The error rate of the adiabatic CNOT gate in the presence of dephasing has been reported [9].

Recently, we have proposed and found [10] pulse sequences for non-adiabatic two-qubit gates (CNOT, swap and controlled Z gates) for the Kane quantum computer architecture. These quantum gates are simpler, with higher fidelity, and faster than existing proposals. Any two-qubit gate may be easily found and implemented using similar pulse sequences. In addition, we have also investigated the effect of decoherence on the operations of these non-adiabatic quantum gates and some simple quantum algorithms for a variety of dephasing rates [11].

6. Spin measurements

The computations of the proposed quantum computer are done when $J < \mu_B B/2$ where the electrons are fully polarized. Measurements are, however, made when $J > \mu_B B/2$ where electron $|a_e\rangle$ states have the lowest energies (see Fig. 4 in Ref. 1). As the electron levels cross (see Fig. 4 in Ref. 1), the $|\downarrow\downarrow\rangle$ and $|a_e\rangle$ states are coupled by hyperfine interactions, Eq. (4), with the nuclei. The no-level crossing theorem in quantum mechanics states that a pair of energy levels connected by perturbation do not cross as the strength of the perturbation is varied. For the lowest eight energy states, only two pairs of states are connected by hyperfine interactions, Eq. (4), as the exchange energy J is varied. The first pair of states consists of $|\downarrow\downarrow\rangle|a_n\rangle$ and $|a_e\rangle|11\rangle$, and the members of the other pair are $|\downarrow\downarrow\rangle|00\rangle$ and $|a_e\rangle|a_n\rangle$ states. The coupled states in each pair hybridize and their energy levels repel each other, leading to anti-crossing behavior in the vicinity of $J = \mu_B B/2$, shown in Fig. 2. Other energy levels, however, do cross each other since no coupling exists between them. Thus during an adiabatic increase in J (from $J < \mu_B B/2$ to $J > \mu_B B/2$), the lowest eight states evolve, as illustrated in Fig. 2, as follows [8]:

$$
\begin{array}{llll}
|a_e\rangle|11\rangle & \longrightarrow & |\downarrow\downarrow\rangle|a_n\rangle, & |\downarrow\downarrow\rangle|11\rangle & \longrightarrow & |\downarrow\downarrow\rangle|11\rangle, \\
|a_e\rangle|a_n\rangle & \longrightarrow & |\downarrow\downarrow\rangle|00\rangle, & |\downarrow\downarrow\rangle|s_n\rangle & \longrightarrow & |\downarrow\downarrow\rangle|s_n\rangle, \\
|a_e\rangle|s_n\rangle & \longrightarrow & |a_e\rangle|s_n\rangle, & |\downarrow\downarrow\rangle|a_n\rangle & \longrightarrow & |a_e\rangle|11\rangle, \\
|a_e\rangle|00\rangle & \longrightarrow & |a_e\rangle|00\rangle, & |\downarrow\downarrow\rangle|00\rangle & \longrightarrow & |a_e\rangle|a_n\rangle.
\end{array}
\tag{10}
$$

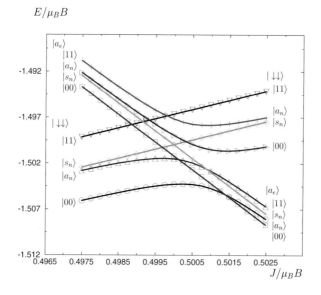

Figure 2.   Adiabatic evolution of the lowest eight energy states in the vicinity of $J = \mu_B B/2$ with $A_1 = A_2 \approx 0.001 \, \mu_B B$.

We now describe how the polarization of a single ^{31}P nuclear spin can be measured using the idea of adiabatic evolution of states. The idea is to transfer the detection of the nuclear spin to the electrons. Suppose that at $J = 0$, $A_1 > A_2$, the energy levels of nuclear spin states in electron spin ground state, $|\downarrow\downarrow\rangle$, are in the following order with $|10\rangle$ above $|01\rangle$ ($|10\rangle$ and $|01\rangle$ are degenerate when $A_1 = A_2$): $|11\rangle$, $|10\rangle$, $|01\rangle$, and $|00\rangle$. This procedure distinguishes energy state of nuclear spin #1 from that of nuclear spin #2. If now J is turned on and increased adiabatically (but still $J < \mu_B B/2$) and at the same time $\Delta A = A_1 - A_2$ is turned off adiabatically, then the $|10\rangle$ and $|01\rangle$ states evolve gradually into $|s_n\rangle$ and $|a_n\rangle$ respectively. The above few steps of operations are similar to the first few steps performed in the controlled-NOT procedure described in Sec. 5. If J is increased further adiabatically from $J < \mu_B B/2$ to $J > \mu_B B/2$, according to Fig. 2 or Eq. (10), the system with nuclear spin state in either $|11\rangle$ or $|s_n\rangle$ will remain in the same electron-nuclear state. On the other hand, the electron spin state of the system with nuclear spin state in either $|00\rangle$ or $|a_n\rangle$ will evolve into the $|a_e\rangle$ state. Thus the initial orientation of nuclear spin #1 alone will determine what electron spin state the system evolves into. To be more evident, if initially the system is in the electron spin polarized $|\downarrow\downarrow\rangle$ state and the orientation of nuclear spin #1 is down, i.e., in $|1\rangle$ state, by means of the above sequence of adiabatic steps, no matter what the orientation of the second nuclear spin has, the system will remain in the same electron spin state $|\downarrow\downarrow\rangle$. However, if initially nuclear spin #1 is in $|0\rangle$ state, the electron spin state of the system will evolve into $|a_e\rangle$ state. We summarize the sequence of steps performed and the associated adiabatic evolution of states in the following:

$$
\begin{array}{cccc}
J=0 & 0<J<(\mu_B B/2) & 0<J<(\mu_B B/2) & J>(\mu_B B/2) \\
A_1 > A_2 & \Delta A \gg h\nu_J & A_1 = A_2 & A_1 = A_2 \\
|\downarrow\downarrow\rangle|11\rangle & |\downarrow\downarrow\rangle|11\rangle & |\downarrow\downarrow\rangle|11\rangle & |\downarrow\downarrow\rangle|11\rangle, \\
|\downarrow\downarrow\rangle|10\rangle \longrightarrow & |\downarrow\downarrow\rangle|10\rangle \longrightarrow & |\downarrow\downarrow\rangle|s_n\rangle \longrightarrow & |\downarrow\downarrow\rangle|s_n\rangle, \\
|\downarrow\downarrow\rangle|01\rangle \longrightarrow & |\downarrow\downarrow\rangle|01\rangle \longrightarrow & |\downarrow\downarrow\rangle|a_n\rangle \longrightarrow & |a_e\rangle|11\rangle, \\
|\downarrow\downarrow\rangle|00\rangle & |\downarrow\downarrow\rangle|00\rangle & |\downarrow\downarrow\rangle|00\rangle & |a_e\rangle|a_n\rangle.
\end{array}
\tag{11}
$$

From Eq. (11), if the final electron spin states, $|\downarrow\downarrow\rangle$ and $|a_e\rangle$, of the two neighboring donor atom system are, by some means, distinguishable from each other, the initial orientation of nuclear spin #1 can thus be determined.

A method [1,8] to detect the nuclear and electron spin states using electronic means is shown in Fig. 3. The basic idea is to turn the spin measurement into a charge measurement. Suppose a two-qubit quantum gate operation was just done. These two qubits located right below the A gates in Fig. 3, have no coupling between them and are now ready to be read out (for the purpose of quantum computing, the qubits must be coupled only for the short time of switching, while most of time there is to be no coupling between them). Let us call the other two donor atoms located below the single-electron transistors (SETs) and on the left and right of the two qubits in Fig. 3 the L-donor and R-donor atoms respectively. For the purpose of nuclear spin readout, we take the L-donor and the neighboring

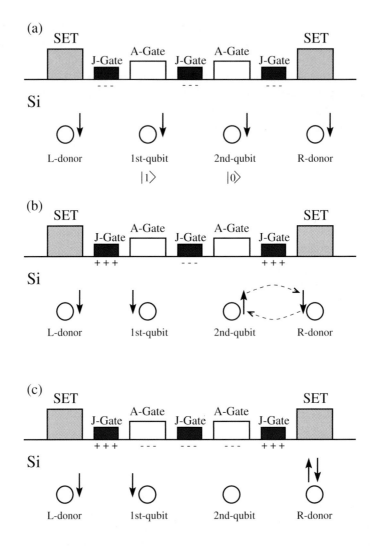

Figure 3.　Schematic illustration of the two-qubit and spin measurement system. (a) The two qubits are initially in nuclear spin $|1\rangle$ and $|0\rangle$ states respectively and all the donor electrons are all in the spin $|\downarrow\rangle$ states. (b) After the sequence of adiabatic process for spin readout stated in Eq. (11), the electron state of the system consisting of the L-donor and first qubit evolves into $|\downarrow\downarrow\rangle$ state, while the electron state of the other system containing the second qubit and the R-donor atom evolves into $|a_e\rangle$ state. (c) By applying appropriate bias of the A gate voltage above the two qubits, only the electrons in $|a_e\rangle$ state in the system consisting of the 2nd qubit and R-donor atom can make transitions into a state in which the two electrons are bound to the same donor (D^- state). The electron current during the transition is measured using highly charge-sensitive SET, enable the underlying spin states of the electron and nuclei to be determined.

first qubit as a system, and the second qubit and its neighboring R-donor atom as the other system. In order to investigate the difference in spin detection between the two possible values of nuclear spin, we assume that the nuclear spin of one of the two qubits, say the first qubit, is in $|1\rangle$ state and the nuclear spin of the other qubit, say the second qubit, is in $|0\rangle$ state. After the sequence of adiabatic steps stated in Eq. (11) (which can be performed simultaneously for both of the systems), the electron state of the system consisting of the L-donor and first qubit will remain in $|\downarrow\downarrow\rangle$ state, and the electron state of the other system containing the second qubit and the R-donor atom will evolve into $|a_e\rangle$ state, illustrated schematically in Fig. 3 (b). The ^{31}P donor in Si has a stable two-electron state (D$^-$ state) with a second electron binding energy of 1.7meV, but only if the relative spin of the two electrons is a singlet. Therefore even if the A gates above the two qubits are biased appropriately, electron charge motion between the donors will only occur if the electrons are in the singlet $|a_e\rangle$ state. For the above assumed initial values for the two qubits, only the electron on the second qubit (initially in $|0\rangle$ state) will tunnel into R-donor atom forming a D^- state. As a result, a perturbation on the conductance of the highly charge-sensitive SET above the R-donor atom will be observed. On the other hand, no change on the conductance of the SET above the L-donor atom is expected. To sum up, the job of measuring spin is converted into a job of measuring charge movement of electron tunneling into D^- state. If a perturbation signal is observed in the SET, the nuclear spin of the measured qubit is in $|0\rangle$; if the SET does not detect any change in tunneling current, then the measured qubit is in $|1\rangle$.

Recently, other alternative proposals to read out or detect the spin state for the Kane quantum computer architecture have been suggested [12,13].

7. Conclusion

We have exploited the silicon-based electron-mediated nuclear spin quantum computer proposed by Kane. Especially, the single- and two-qubit energy levels and eigen states as well as their operations and measurements have been discussed in details. A brief up-to-date development and progress on the Kane quantum computer have also been described. The author would like to thank C.D. Hill, L.M. Kettle and G.J. Milburn for their helpful discussions and research collaborations related to this work.

References

1. B. E. Kane, Nature **393**, 133 (1998).
2. B. E. Kane, unpublished manuscript (1998).

3. R.G. Clark, R. Brenner, T.M. Buehler, V. Chan, N.J. Curson, A.S. Dzurak, E. Gauja, H.-S. Goan et al., Phil. Trans. R. Soc. of Lond. A **361**, 1451 (2003).
4. P. Y. Yu and M. Cardona, *Fundamentals of Semiconductors* (Springer-Verlag, Berlin, 1996).
5. L.M. Kettle, H.-S. Goan, S.C. Smith, C.J. Wellard, L.C.L. Hollenberg and C.I. Pakes, Phys. Rev. B **68**, 075317 (2003); L.M. Kettle, H.-S. Goan, S.C. Smith, L.C.L. Hollenberg and C.J. Wellard, , J. Phys: Condensed Matter **16**, 1011 (2004).
6. B. Koiller et al., Phys. Rev. Lett. **88**, 027903 (2002).
7. C.J. Wellard, L.C.L. Hollenberg, F. Parisoli, L.M. Kettle, H.-S. Goan, J.A.L. McIntosh and D.N. Jamieson, Phys. Rev. B **68**, 195209 (2003); C.J. Wellard, L.C.L. Hollenberg, L.M. Kettle, and H.-S. Goan, J. Phys: Condensed Matter **16**, 5697 (2004).
8. H.-S. Goan and G.J. Milburn, unpublished manuscript (2000).
9. A.G. Fowler et al., Phys. Rev. A **67**, 012301 (2003).
10. C.D. Hill and H.-S. Goan, Phys. Rev. A **68**, 012321 (2003).
11. C.D. Hill and H.-S. Goan, Phys. Rev. A **70**, 022310 (2004); C.D. Hill and H.-S. Goan, AIP Conference Proceedings **734**, 167 (2004).
12. T.A. Brun and H.-S. Goan, Phys. Rev. A **68**, 032301 (2003).
13. L.C.L. Hollenberg et al. Phys. Rev. B **69**, 233301 (2004); A.D. Greentree et al., Phys. Rev. B **70**, 041305 (2004).

ROBUST DYNAMICAL DECOUPLING: FEEDBACK-FREE ERROR CORRECTION*

D. A. LIDAR

*Chemical Physics Theory Group, Chemistry Department, and
Center for Quantum Information and Quantum Control, University of Toronto,
80 St. George St., Toronto, ON M5S 3H6, Canada
E-mail: dlidar@chem.utoronto.ca*

K. KHODJASTEH

*Department of Physics, University of Toronto,
60 St. George St., Toronto, ON M5S 1A7, Canada
E-mail: kaveh@physics.utoronto.ca*

Dynamical decoupling is a feed-back free scheme for quantum error correction against noise and decoherence errors. An efficiency analysis of dynamical decoupling is performed. Furthermore we provide the basic concepts of dynamical decoupling and quantum error correcting codes, and give an example of a hybrid protection scheme. Some interesting extensions of dynamical decoupling are discussed at the end.

1. Introduction

Recent advances in control and measurements of quantum systems have given rise to the exciting field of quantum information processing. In quantum information processing different aspects and ideas from quantum physics, computer science, and mathematics have promised futuristic technologies such as quantum computing (QC) and secure quantum communications[11]. The starting point for these application is usually a collection of separate, yet jointly controllable quantum systems, such as qubits. A universal set of the possible dynamics on the system allows for producing arbitrary unitary operators on the full Hilbert space of these qubits. These control operations are sometimes referred to as quantum gates in a discrete sense, or one may alternatively use a set of control Hamiltonians to achieve universality. Quantum information processing has other important

*Financial support from the DARPA-QuIST program (managed by AFOSR under agreement No.F49620-01-1-0468), the Sloan foundation, and PREA is gratefully acknowledged.

existential requirements such as efficient state measurement and state preparation.

Isolation of the qubits of a quantum computer is a mathematical simplification. In reality every quantum system may interact with others. This interaction and lack of information about the state of those other systems is a major obstacle in quantum information processing. In other words, realistic and arbitrary control of quantum systems is always limited by the problem of decoherence and noise. These problems are associated with the undesired interactions of an ideally isolated quantum system, over which some means of manipulation and control already exists. More precisely, decoherence refers to the decay of *quantum* superpositions of pure states into mixed states of the possible measurement outcomes[11]. The term "error" often takes a more general meaning as it applies to classical circumstances also.

In the context of quantum information processing, "error correction" theory discusses problems (and solutions) due to the undesired interactions of quantum systems. Sometimes some knowledge of the error process is available and can be used to protect quantum systems against errors. For example, the collective decoherence models where certain global symmetries exist in the error process can easily be dealt with encoding of quantum information in the available decoherence free subspaces[10]. Universal error correction, in contrast, focuses on more general error models where apart from certain error rates and error correlations, not much is assumed about the nature of the errors affecting the system. Different regimes of errors naturally demand different error correction schemes. Practically speaking, these schemes can be operationally active or passive, use feedback, or use a larger operational Hilbert space (encoding). The domain of applicability and practicality of various schemes are different and so far no single scheme is practically capable of protecting against arbitrary error types/models. Despite this lack of generality, it is widely believed that hybrid methods incorporating various schemes can be efficiently used within a given physical implementation framework of a quantum information processor, for the purpose of quantum computing, and quantum state preservation [8,9].

In this summary we shall refer to the system's Hilbert space as $\mathcal{H}_S$, and to that of an external environment as $\mathcal{H}_B$, which we shall refer to as "the bath" or "the environment". We often assume $\mathcal{H}_S$ to be composed of one or more qubits. Generically ψ refers to a pure quantum state, while ρ refers to a density matrix. The starting point of the analysis in quantum error correction theory is the system-bath Hamitonian ansatz:

$$H = I_B \otimes H_S + H_B \otimes I_S + \underbrace{\sum_\alpha B_\alpha \otimes S_\alpha}_{H_{SB}} \tag{1}$$

Operators I_B and I_S are the identity operators on $\mathcal{H}_S$ and $\mathcal{H}_B$. Typically no reliable operation or observation can be made on the environment. H_S refers to the system Hamiltonian, over which certain control is assumed. H_B is the Hamiltonian for the environment which might be unknown. H_{SB} is responsible for entangling the states of the system and the environment, which leads to decoherence, once the system density matrix ρ_S is reproduced after tracing out the environment: $\rho_S = \text{Tr}_B(\rho_{SB})$. The isolation and the assumption of control over the system is a basic requirement of quantum information processing, but Eq. (1) is an approximation that can always be improved by incorporating further entities within the environment.

An alternative description involving only the system is also used: A quantum channel describes the linear transformation of the system density matrix ρ_S. This description in the Markovian regime is further simplified in the sense that the quantum channel description for short times describes the evolution of the system for all times and the whole evolution can be generated by exponentiating a "Lindblad super-operator" acting on the density matrices[1]. It is worth mentioning that while these pictures are mathematically interchangeable, the physical constraints often limit the way these pictures are used. Quantum error correcting codes[8] and decoherence free subspaces as error correction protocols can be described in the channel picture. Quantum error correcting codes in particular are the most predominant error correction strategy as they offer extensive universality within the Markovian regime and allow for fault tolerant quantum computation which technically refers to a robust implementation of quantum computing (in contrast with quantum state preservation or quantum memory). Quantum error correction carries the overhead of extra quantum computing qubits, entangling gates, measurements and ancillary qubits.

Dynamical Decoupling[17] is another error correction strategy which is implemented by application of a series of fast and strong/narrow pulses acting on the system that effectively renormalize the interaction Hamiltonian to remove undesired terms such as H_{SB}. Dynamical decoupling techniques have been traditionally used in NMR to remove unwanted intra-nuclear couplings and obtain high resolution spectra[4] . It can be shown that dynamical decoupling can approximately remove an arbitrary H_{SB} Hamiltonian, a technique which is referred to as universal dynamical decoupling. While this universality is a remarkable aspect of the dynamical decoupling theory, comparable to the universality of quantum error correcting codes, physical constraints have obscured the prospect of effectively using dynamical decoupling in quantum information processing. These constraints are the requirements for perfect pulses and the ability to run closely packed sequence of pulses.

In this summary we analyze and discuss dynamical decoupling in the Hamiltonian setting, and sketch some interesting extensions of dynamical decoupling. We also cover the basics of quantum error correcting codes and present an example of a hybrid error correction scheme where both dynamical decoupling and quantum error correcting codes are used.

2. Propagators, Pulses, and Idealizations

In this summary we shall only focus on qubit systems. Some of the results and ideas that apply to the qubit case are extendible to other systems and settings. An isolated qubit is always driven by an $su(2)$ Hamiltonian given by $H_c = h_x X + h_y Y + h_z Z$. Operators X, Y, and Z refer to the corresponding Pauli operators. Without loss of generality, all Hamiltonian components which we shall consider are either traceless or a multiple of identity.

The unitary operations on this system can be generated by the Schrödinger propagator between the times t_0 and t_1:

$$U = T_+ \left[\exp(-i \int_{t_0}^{t_1} H_c(t')dt') \right] \qquad (2)$$

For example the unitary operator X can be generated by turning on $H_c = \pi/2h_x X$ for a duration $\delta = t_1 - t_0 = \frac{1}{h_x}$. We shall refer to $\delta \to 0$ as an "ideal pulse" in this summary. If a unitary operator is given by $e^{-i\Phi}$, we will simply refer to the Hermitian operator Φ as the *phase*.

As discussed in the previous section, a general Hamiltonian describing the qubit S plus an environment B is given by Eq. (1), and in the qubit case can be generally written as

$$H = H_c + H_e \qquad (3)$$
$$= I_B \otimes (h_x X + h_y Y + h_z Z) + B_x \otimes X + B_y \otimes Y + B_z \otimes Z + B_0 \otimes I_S \qquad (4)$$

where H_e loosely refers to the undesired parts of the qubit (and the environment) Hamiltonian. In this picture the propagator for a given navigation of the control H_c is

$$U_e = T_+ \left[\exp(-i \int_{t_0}^{t_1} (H_c(t') + H_e)dt') \right] \qquad (5)$$

By scaling down $t_1 - t_0$ to 0 while scaling up $H_c(t)$ to keep the product constant, we get the ideal propagator: $U_e = I_B \otimes U$. In most of what follows we assume either ideal pulses or rectangular pulses.

3. Dynamical Decoupling

For the sake of clarity we first focus on the simplest case which we refer to as a "canonical dynamical decoupling cycle". We start from Eq. (3) and restrict $H_e = B_z \otimes Z$. As in the previous section H_c is the controllable part of the Hamiltonian and we use it to produce unitary pulses acting on the system. The *canonical dynamical decoupling cycle with pulse X* is given by "$X F_{[\tau]} X F_{[\tau]}$", where the pulse sequence is applied from the right and $F_{[\tau]}$ refers to a free evolution of duration τ. The unitary propagator for this sequence is given by

$$
\begin{aligned}
U &= X \exp[-i\tau(B_z \otimes Z)] X \exp[-i\tau(B_z \otimes Z)] \\
&= \exp[-i\tau(B_z \otimes XZX)] \exp[-i\tau(B_z \otimes Z)] \\
&= \exp[-i\tau(-B_z \otimes Z)] \exp[-i\tau(B_z \otimes Z)] = I_S \otimes I_B
\end{aligned}
\tag{6}
$$

The above sequence has removed H_e from the evolution of the system by *time reversal*. To generalize we note that any qubit Hamiltonian H_e can be decomposed as $H_e = H_e^{X,\|} + H_e^{X,\perp}$ such that $[X, H_e^{X,\|}] = 0$ and $\{X, H_e^{X,\perp}\} = 0$. For a general H_e we can rewrite the above sequence

$$
\begin{aligned}
U &= X \exp[-i\tau(H_e^{X,\|} + H_e^{X,\perp})] X \exp[-i\tau(H_e^{X,\|} + H_e^{X,\perp})] \\
&= \exp[-i\tau(H_e^{X,\|} - H_e^{X,\perp})] \exp[-i\tau(H_e^{X,\|} + H_e^{X,\perp})] \\
&=: \exp(-i2\tau \mathcal{D}_X[H_e]) = \exp[-i\tau(2H_e^{X,\|} + O(B_\alpha^2 \tau)]
\end{aligned}
\tag{7}
$$

In Eq. (7) the overall propagator is used to define an effective Hamiltonian $\mathcal{D}_X[H_e]$. We can look at the transformation of the propagator as a renormalization of the Hamiltonian:

$$
B_x \otimes X + B_z \otimes Z + B_Y \otimes Y \xmapsto{X F_{[\tau]} X F_{[\tau]}} B_x \otimes X + O(B_\alpha^2 \tau)
\tag{8}
$$

Geometrically this process can be thought of a projection of H_e parallel to the "X axis".

The pulse used in the canonical dynamical decoupling sequence, removes all terms anti-commuting with it from the effective Hamiltonian. To remove every possible term it suffices (for example) to use the propagator of the above sequence as the "free evolution period" of a Y canonical dynamical decoupling sequence:

$$
Y F_{[2\tau]} Y F_{[2\tau]} \curvearrowright Y X F_{[\tau]} X F_{[\tau]} Y X F_{[\tau]} X F_{[\tau]}
\tag{9}
$$

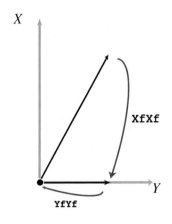
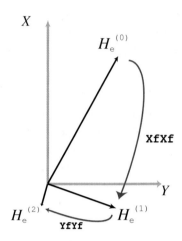

Figure 1.   Projections and rotated projections in dynamical decoupling. The aim is to produce a combination of two orthogonal projectors (left), but in practice due to non-commutative terms, we obtain a combination of two tilted projectors.

Optionally we can use $YX = Z$ to simplify[a] the above sequence: $ZF_{[\tau]}XF_{[\tau]}ZF_{[\tau]}XF_{[\tau]}$. This sequence is now *universal* in the sense that any H_e implied in the free evolution period $F_{[\tau]}$ is removed up to $O(B_\alpha^2\tau)$ since the only terms that commute with both X and Y are the "pure-bath" operators that have no effect on the system dynamics. Geometrically we can represent the combination of the two above transformations as two projections on the space of Hamiltonians. Ideally these two orthogonal projections will produce zero, but the higher order terms can be shown to produce extra rotations that suppress the cancellation and generally result in imperfect decoupling. This is illustrated in Fig. 1.

The unitary propagator of the sequence of sign-flipped Hamiltonians can be approximated by a Magnus expansion[17]: If the Hamiltonian for the Schrödinger evolution of a quantum system is given by $H(t)$, then the propagator $U(t)$ from 0 to t is given by $U = \exp(A_1 + A_2 + \ldots)$, where

$$A_1 = i \int_0^t dt_1 H(t_1) \tag{10}$$

$$A_2 = \frac{1}{2} \int_0^t dt_1 \int_0^{t_1} dt_2 [H(t_2), H(t_1)] \tag{11}$$

and A_i for $i > 2$ are more complicated *i*-dimensional integrals of commutators involving $H(t)$s at *i* times. A *k-th order universal dynamical decoupling sequence*

---

[a]We omit global phase factors of i, -1, *etc.*

produces a sequence of Hamiltonians and intervals such that A_1 to A_k contain only pure-bath terms $B \otimes I_S$. The sequence $ZF_{[\tau]}XF_{[\tau]}ZF_{[\tau]}XF_{[\tau]}$ is, for example, 1st order universal. If we can bound the operators B_α form above by some error rate j, then the second order Magnus terms for this sequence will be bounded by $c\tau_0^2 j^2$, where c is a numerical constant of the order of 10. The Magnus terms can be used to produce an effective "phase" or Hamiltonian for imperfections of dynamical decoupling.

In periodic dynamical decoupling, N periodic repetitions of a universal sequence for a total duration of $T = 4N\tau$ are used. The leading error associated with this decoupling can be easily estimated:

$$\|\Phi\| \lesssim Nc\tau_0^2 j^2 = c(Tj)^2/N. \tag{12}$$

The actual error in the fidelity is typically given by the square of this phase [15].

Eq. (12) dictates the use of shorter times between the pulses to obtain higher fidelities, however once more pulses are used, realistic "per pulse errors" take over and produce significant extra errors not covered by the above formula. It should be noted that even perfect control over H_c does not guarantee perfect pulses, as the presence of the H_e terms in the Hamiltonian, will result in systematic pulse errors proportional to the pulse width[b] and once the total duration for a physical realization of the sequence is fixed, one cannot make τ_0 arbitrary small by including increasingly many finite width pulses.

Despite these difficulties dynamical decoupling has been used with great success in the context of NMR. It is worth emphasizing that three main parameters enter the analysis of dynamical decoupling: model considerations, system-bath couplings, and pulse imperfections. These parameters will also appear in the analysis of quantum error correcting codes.

4. Basic Quantum Error Correcting Codes

Based on successful methods from "classical" error correcting codes, quantum error correcting codes are thought to be the most generic and best understood error protection schemes. The starting point is to obtain the error operators E_α, that describe the evolution of an ideally fixed (at $\rho(0)$) density matrix for the system via a Kraus operator expansion:

$$\rho(t) = \sum_\alpha E_\alpha(t)\rho E_\alpha(t)^\dagger \tag{13}$$

---

[b]It is possible to perform dynamical decoupling without the requirement of infinitely sharp pulses [16]. Nonetheless strong pulses and precise control are required to achieve high fidelities.

Error correcting codes are efficient in the Markovian regime where the correlations in time are minimal, and the short time behavior of the system dictates (E_α instead of $E_\alpha(t)$) that of the longer times. Quantum error correcting codes encode (embed) "logical qubit" states into a larger Hilbert space (of many more qubits). The embedding is such that the effect of all error operators E_α on all encoded states $|i\rangle_L$ are mutually orthogonal:

$$_L\langle i| E_\alpha^\dagger E_\beta |j\rangle_L = \delta_{ij} c_{\alpha\beta} \tag{14}$$

where c_{ab} are the elements of a Hermitian matrix. This ensures that the effect of each error on any state in the logical Hilbert space can be measured via a "syndrome measurement" protocol and the proper state can be obtained via a "recovery" protocol. Structurally quantum error correcting codes rely on successive measurements, near parallelism in the applied operations, enlargement of the qubit Hilbert space[c] and availability of refreshable, pure ancilla qubits. Quantum error correcting codes are most efficient for the independent noise model, in which qubits are acted upon by errors individually and randomly such that there is no correlation in time (successive errors) and space (multiple qubit errors). For this case the theory can be applied fault-tolerantly together with actual encoded quantum operations. Within this theory one can show that the same embedding of the logical states in a bigger Hilbert space can be applied recursively so that errors, with rates lower than a certain threshold, can be efficiently removed. This is what is known as the threshold theorem for concatenated error correcting codes: For error rates below a certain threshold, using an exponential number of resources (qubits, pulses, and measurements) leads to super-exponential improvement in performance. Different assumptions on error model and modifying the actual error correction/computation schemes results in different thresholds [14,7].

It is possible to combine quantum error correcting codes with quantum operations. The stabilizer theory of the quantum error correcting codes[3] provides a relatively simple way of embedding the universal quantum operations with a given stabilizer based quantum error correcting code, and is the basis of fault-tolerant quantum computation.

The stabilizer formalism relates to dynamical decoupling also: It can be shown that the stabilizer elements of a given quantum error correcting code can be used as canonical decoupling cycles on the physical qubits of that code for removing the error operators that the code can correct, from the interaction terms. In other words the stabilizer generator elements become the canonical dynamical decoupling pulses; *e.g.*, for a single qubit the stabilizer generators can be taken to be X

---

[c]The smallest code that can correct arbitrary 1-qubit errors is given in term of 5-qubit states.

and Y.

5. Hybrid Error Correction Schemes

While both dynamical decoupling and quantum error correcting codes work for arbitrary error types, they each depend on certain resources and have their domain of practicality. However, certain modes of decoherence might allow for different techniques to be used together. One such example is the error correction of spontaneous emission on an electronic-level qubit. The coupling of a bound electron (system) with the electromagnetic field (environment) results in spontaneous emission, in which the electron that might be in a superposition of the excited ($|1\rangle$) and the ground ($|0\rangle$) state decays to $|0\rangle$, and the quantum information stored in the superposition is lost. Assume a collection of such electron-level qubits labeled by $i = 1,\ldots,n$. Evolution of the system can be described in the quantum trajectories picture. In this picture, each trajectory is separated into non-unitary evolution intervals, interrupted by sudden application of error operators (E_i) ($E_i = |0\rangle_i \langle 1|$ for spontaneous emission on the qubit i). The final density matrices of all trajectories can then be averaged to give the actual probabilistic density matrix as a function of time. The non-unitary evolution in this picture is generated by a non-Hermitian conditional Hamiltonian, $H_{\text{cond}} = H_{\text{system}} - \frac{i}{2} \sum_i E_i^\dagger E_i$. In the case of spontaneous emission we obtain $H_{\text{cond}} = i \sum_i |1\rangle_i \langle 1|$. We showed[5] that a simple quantum error correcting code involving only one extra qubit together with a dynamical decoupling pulses sequence with pulse type X can be used to correct spontaneous emission errors as long as the error rate is small and the pulse operations are faster than the average time between the errors. We further investigated the possibilities of fault-tolerant quantum computation within this setting.

6. Extensions of Dynamical Decoupling

To conclude, in the following subsections we briefly review some of the most recent results and ideas in dynamical decoupling.

6.1. *Inter-qubit couplings*

Consider a collection of n qubits in which the undesired interactions not only involve the couplings with the environment, but also inter-qubit couplings such as nearest neighbor couplings terms: $\sigma_x^i \sigma_x^j$. An interesting extension of dynamical decoupling is to remove these error terms, along with the system-bath interaction. The optimization of this sequence in terms of the number of pulses required becomes a combinatorial optimization problem[12].

6.2. *Zeno Effect*

On a different note an interesting connection is observed between dynamical decoupling and the quantum Zeno effect which traditionally is understood in terms of repeated measurements on a quantum system to inhibit or enhance couplings. This traditional view has been transformed to include arbitrary quantum operations and naturally includes dynamical decoupling as a special case [2]. Furthermore recently it has been shown[13] that quantum error correcting codes can be defined in terms of the Zeno effect, which hints to a possibly deeper physical connection between different error correction schemes.

6.3. *Concatenation*

The idea of concatenation, *i.e*, the recursive embedding of encoded qubits in the same error correcting code, has been used for threshold calculations. This threshold refers to an initial error rate below which, even after using an exponential number of concatenated qubits, the final error rate is inhibited "super-exponentially". We showed[6] that the same idea can be used with dynamical decoupling with a similar threshold. Given the number of pulses used N, the phase associated with concatenated dynamical decoupling scales as

$$\|\Phi\| \lesssim \frac{1}{c}(cj/N)^N. \tag{15}$$

This should be contrasted with the result obtained in the case of periodic decoupling, Eq. (12). The exponential improvement in terms of the number of pulses required thus strongly suggests the use of concatenated pulse sequences over the periodic schemes.

6.4. *Error per Gate*

One of the main difficulties with both dynamical decoupling and quantum error correcting codes is the requirement for near perfect quantum gates. For example, consider an error correcting code designed to reduce the rate e to some lower error rate e^2. In fault-tolerance theory, however, the error per gate is another important factor. Especially when a high number of quantum gates are executed, one needs to make sure that the gate errors are still corrected with the original error correcting design. Now, suppose the errors are based on a continuous model such as a Hamiltonian picture. In this picture we may associate linear error accumulation (well-defined error rate) to short times. Now one might ask "is the error probability fundamentally different when there are gates acting on the system in comparison to when no gate is being applied?" If the error rates for "gate-free" and "gate" evolution are of the same order, then there is effectively only one important error rate in the system. Define the gate-effective error generator H'_e as $\exp(-i\delta H'_e) = \exp(-i\delta(H_e + H_P))\exp(i\delta H_P)$, where $\exp(-i\delta H_P)$ is the propagator for a pulse P, and H_e is the free evolution error generator. For the case where all operators belong to $su(2)$, we have shown that $\|H'_e\| \leq \|H_e\|$. Thus for a simple case we confirm that the natural error parameter for the free evolution of the system is indeed larger than the parameters that are associated with gate errors, which simplifies fault tolerance.

6.5. *Dynamically Decoupled Quantum Computing*

While dynamical decoupling is an essential part of engineering interactions for quantum computing in many proposals, a *universal* decoupling process will naturally remove the "desired" evolution of the system along with the undesired parts. Due to this, the prospects of combining dynamical decoupling with quantum computation have been limited to a scenario where dynamical decoupling stops so that a quantum gate or a measurement is performed and is then resumed. This method is obviously prone to errors accumulating during the computation phase and the advantages of dynamical decoupling with state preservation are rendered useless. Nonetheless a protocol can be constructed in which an encoding of a few qubits are employed so that the quantum computing operations can be embedded along with the dynamical decoupling pulse sequence which corrects for single qubit errors. In this construction dynamical decoupling reflects and brings back all trajectories of the qubits that leave the encoded subspaces without modifying the desired encoded dynamics.

References

1. H.J Carmichael. *An Open Systems Approach to Quantum Optics.* Springer-Verlag, Berlin, 1993.
2. P. Facchi, D. A. Lidar, and S. Pascazio. Unification of dynamical decoupling and the quantum zeno effect. *Phys. Rev. A*, 69(3):032314, 2004.
3. D. Gottesman. Theory of fault-tolerant quantum computation. *Phys. Rev. A*, 57:127, 1998.
4. U. Haeberlen. *High Resolution NMR in Solids: Selective Averaging.* Academic Press, 1976.
5. K. Khodjasteh and D. A. Lidar. Quantum computing in the presence of spontaneous emission by a combined dynamical decoupling and quantum-error-correction strategy. *Phys. Rev. A*, 68(2):022322, 2003.
6. Kaveh Khodjasteh and Danbiel A. Lidar. Concatenated dynamical decoupling. *e-print:* `http://arxiv.org/abs/quant-ph/0408128`, 2004.

7. E. Knill. Fault-tolerant postselected quantum computation: Threshold analysis. *e-print:* `http://arxiv.org/abs/quant-ph/0404104`, 2004.
8. E. Knill and R. Laflamme. Theory of quantum error-correcting codes. *Phys. Rev. A,* 55:900, 1997.
9. E. Knill, R. Laflamme, and W. Zurek. Resilient quantum computation. *Science,* 279:342, 1998.
10. D.A. Lidar, I.L. Chuang, and K.B. Whaley. Decoherence-free subspaces for quantum computation. *Phys. Rev. Lett.* 81:2594, 1998.
11. M. A. Nielsen and I. L. Chuang. *Quantum Computation and Quantum Information.* Cambridge University Press, 2000.
12. M. Roetteler and P. Wocjan. Equivalence of decoupling schemes and orthogonal arrays. *e-print:* `http://arxiv.org/abs/quant-ph/0409135`, 2004.
13. M. Sarovar and G. J. Milburn. Continuous quantum error correction. *e-print:* `http://arxiv.org/abs/quant-ph/0501049`, 2005.
14. A. Steane. Overhead and noise threshold of fault-tolerant quantum error correction. *Phys. Rev. A,* 68:042322, 2003.
15. B. M. Terhal and G. Burkard. Fault-tolerant quantum computation for local non-markovian noise. *Phys. Rev. A,* 71(1):012336, 2005.
16. L. Viola and E. Knill. Robust dynamical decoupling of quantum systems with bounded controls. *Phys. Rev. Lett.,* 90:037901, 2003.
17. L. Viola, E. Knill, and S. Lloyd. Dynamical decoupling of open quantum systems. *Phys. Rev. Lett.,* 82:2417, 1999.

A BELL INEQUALITY BASED ON CORRELATION FUNCTIONS FOR THREE QUBITS

CHUNFENG WU

Department of Physics, National University of Singapore, 2 Science Drive 3, Singapore 117542

JING-LING CHEN

Department of Physics, National University of Singapore, 2 Science Drive 3, Singapore 117542

L. C. KWEK

Department of Physics, National University of Singapore, 2 Science Drive 3, Singapore 117542

Nanyang Technological University, National Institute of Education, 1, Nanyang Walk, Singapore 637616

C. H. OH[*]

Department of Physics, National University of Singapore, 2 Science Drive 3, Singapore 117542

We discuss a Bell inequality based on correlation function for three qubits. The inequality is violated by any pure entangled state for 3 qubits. The strength of the violation is the same as the result in Ref.[13].

1. Introduction

The gedanken experiment involving the position and momentum of entangled particles proposed by A. Einstein, B. Podolsky and N. Rosen (EPR) [1] in a classic 1935 paper showed that quantum mechanics is an incomplete theory. The incomplete description, they argued, could be avoided by postulating the existence of hidden variables that allow for deterministic predictions. In 1964, based on the

---

[*]Electronic address: phyohch@nus.edu.sg

entanglement of spin-1/2 particles introduced by Bohm [2], John Bell showed that the assumption of local realism had experimental consequences: local realism implies experimentally variable constraints on the statistical measurements of two or more physically separated systems. These constraints, called Bell inequalities, which are satisfied by hidden variable theories, can be violated by the predictions of quantum mechanics. Thus, Bell inequalities made it possible for the first time to distinguish experimentally between a local HV model and quantum mechanics.

Quantifying entanglement based on Bell inequality is an important issue in quantum information theory. In 1991, Gisin [4] demonstrated that every pure bipartite entangled state violates the CHSH-Bell inequality. This was known subsequently as Gisin's theorem and it was probably the first step towards quantifying entanglement. A few years later, the Horodecki family [5] and Werner [6] showed that the CHSH-Bell inequality was insufficient to characterize entanglement of mixed states. This was confirmed subsequently by two independent teams [7] [8], who proposed the most general Bell inequalities (correlation function) for N-qubit with two settings per site. However, Gisin and Scarani [9] noticed that there exist pure states of N-qubit that do not violate any of the inequalities. These states are the generalized Greenberger-Horne-Zeilinger (GHZ) states given by

$$|\psi\rangle_{GHZ} = \cos\xi|0\cdots 0\rangle + \sin\xi|1\cdots 1\rangle \tag{1}$$

with $0 \leq \xi \leq \pi/4$. The GHZ states [10] are for $\xi = \pi/4$. In 2001, Scarani and Gisin noticed that for $\sin 2\xi \leq 1/\sqrt{2^{N-1}}$ the states (1) do not even violate the Mermin-Ardehali-Belinskii-Klyshko (MABK) inequalities [11]. These results prompted Scarani and Gisin to note that "this analysis suggests that MK (in Ref. [12], MABK) inequalities, and more generally the family of Bell's inequalities with two observables per qubit, may not be the 'natural' generalizations of the CHSH inequality to more than two qubits" [9], where CHSH stands for Clauser-Horne-Shimony-Holt.

Note that in Ref. [7] Żukowski and Brukner (ŻB) have derived a general Bell inequality for correlation functions for N qubits. The ŻB inequalities include MABK inequalities as special cases. Ref. [12] shows that (a) For N = even, although the generalized GHZ state (1) does not violate MABK inequalities, it violates the ŻB inequality and (b) For $\sin 2\xi \leq 1/\sqrt{2^{N-1}}$ and N = odd, the correlations between measurements on qubits in the generalized GHZ state (1) satisfy all Bell inequalities for correlation functions, which involve two dichotomic observables per local measurement station.

It therefore appears that Gisin's theorem is not valid for 3-qubit. Recently, we provided a further twist to the results. We constructed a Bell inequality based on probability (i.e. not one based on correlation) [13] and showed that Gisin's theorem holds for 3-qubit systems. In this paper, a Bell inequality involving correlation functions on three qubits is developed. We show that the inequality is violated by quantum mechanics. The violation is the same as that predicted in Ref.[13].

2. A Bell Inequality involving Correlation Functions for 3 Qubits

Consider 3 observers, Alice, Bob and Charlie. Suppose they are each allowed to choose between two dichotomic observables, determined by some local parameters denoted here $\hat{n}_1$ and $\hat{n}_2$. Each observer can choose independently two arbitrary directions. The assumption of local realism implies the existence of two numbers $X(\hat{n}_1)$ and $X(\hat{n}_2)$ (with $X = A, B, C$) each taking values $+1$ and -1, depending on the measurement by the corresponding observable defined by $\hat{n}_1$ and $\hat{n}_2$. In a specific run of the experiment the correlations between all 3 observers can be represented by the product $A(\hat{n}_i)B(\hat{n}_j)C(\hat{n}_k)$, where $i, j, k = 1, 2$. For convenience, we write $A(\hat{n}_i)B(\hat{n}_j)C(\hat{n}_k)$ as $A_iB_jC_k$. The correlation function, in the case of a local realistic theory, is then the average over many runs of the experiment

$$Q(A_i, B_j, C_k) = \langle A(\hat{n}_i)B(\hat{n}_j)C(\hat{n}_k)\rangle = \langle A_iB_jC_k\rangle \tag{2}$$

Similarly, the correlation functions between any two observers can be given

$$Q(A_i, B_j) = \langle A(\hat{n}_i)B(\hat{n}_j)\rangle = \langle A_iB_j\rangle$$
$$Q(A_i, C_k) = \langle A(\hat{n}_i)C(\hat{n}_k)\rangle = \langle A_iC_k\rangle$$
$$Q(B_j, C_k) = \langle B(\hat{n}_j)C(\hat{n}_k)\rangle = \langle B_jC_k\rangle \tag{3}$$

The following inequality holds for the predetermined results:

$$\begin{aligned}
&Q(A_1B_1C_1) - Q(A_1B_2C_2) - Q(A_2B_1C_2) - Q(A_2B_2C_1) \\
&+2Q(A_2B_2C_2) - Q(A_1B_1) - Q(A_1B_2) - Q(A_2B_1) \\
&-Q(A_2B_2) + Q(A_1C_1) + Q(A_1C_2) + Q(A_2C_1) + Q(A_2C_2) \\
&+Q(B_1C_1) + Q(B_1C_2) + Q(B_2C_1) + Q(B_2C_2) \qquad \leq 4
\end{aligned} \tag{4}$$

The proof consists of enumerating all the possible values of $A_i, B_j, C_k (i, j, k = 1, 2)$. This proof is most easily seen by fixing values of A_1, B_1, C_1. We now consider different cases depending on the signs of A_1, B_1, C_1.

1. A_1, B_1, C_1 are all $+1$. The inequality (4) can be written as $(A_2B_2 + 1)(C_2 - 1) \leq 0$. Since $A_2, B_2, C_2 = \pm 1$, $A_2B_2 + 1 = 2$ or 0 and $C_2 - 1 = 0$ or -2, thus the inequality is satisfied regardless of the values A_2, B_2, C_2.

2. A_1, B_1, C_1 are all -1. The inequality (4) can be written as $[B_2(A_2 + 1) + (A_2 - 1)]C_2 \leq 2$. If $A_2 = 1$, one finds $[B_2(A_2 + 1) + (A_2 - 1)]C_2 = 2B_2C_2$ which is not greater than 2 since $B_2, C_2 = \pm 1$. If $A_2 = -1$, one finds $[B_2(A_2 + 1) + (A_2 - 1)]C_2 = -2C_2$ which is not greater than 2 since $C_2 = \pm 1$.

3. $A_1 = 1, B_1 = C_1 = -1$. The inequality (4) can be written as $(A_2C_2 - 1)(B_2 + 1) \le 0$. Since $A_2, B_2, C_2 = \pm 1$, $A_2C_2 - 1 = 0$ or -2 and $B_2 + 1 = 2$ or 0, thus the inequality is satisfied no matter which values A_2, B_2, C_2 take.

4. $B_1 = 1, A_1 = C_1 = -1$. The inequality (4) can be written as $(B_2C_2 - 1)(A_2 + 1) \le 0$. Since $A_2, B_2, C_2 = \pm 1$, $B_2C_2 - 1 = 0$ or -2 and $A_2 + 1 = 2$ or 0, thus the inequality is satisfied no matter which values A_2, B_2, C_2 take.

5. $C_1 = 1, A_1 = B_1 = -1$. The inequality (4) can be written as $A_2B_2(C_2 - 1) + (A_2 + B_2)(C_2 + 1) - C_2 \le 3$. If $C_2 = 1$, one finds $A_2B_2(C_2 - 1) + (A_2 + B_2)(C_2 + 1) - C_2 = 2(A_2 + B_2) - 1$ which is not greater than 3 since $A_2, B_2 = \pm 1$. If $C_2 = -1$, one finds $A_2B_2(C_2 - 1) + (A_2 + B_2)(C_2 + 1) - C_2 = -2A_2B_2 + 1$ which is not greater than 3 since $A_2, B_2 = \pm 1$.

6. $A_1 = B_1 = 1, C_1 = -1,$. The inequality (4) can be written as $A_2B_2C_2 - A_2 - B_2 + C_2 \le 4$. If $A_2 = 1$, one finds $A_2B_2C_2 - A_2 - B_2 + C_2 = (B_2 + 1)(C_2 - 1)$ which is less than 4 since $B_2, C_2 = \pm 1$. If $A_2 = -1$, one finds $A_2B_2C_2 - A_2 - B_2 + C_2 = (1 - B_2)(C_2 + 1)$ which is not greater than 4 since $A_2, B_2 = \pm 1$.

7. $A_1 = C_1 = 1, B_1 = -1$. The inequality (4) can be written as $A_2B_2(C_2 - 1) + A_2(C_2 + 1) \le 2$. If $C_2 = 1$, one finds $A_2B_2(C_2 - 1) + A_2(C_2 + 1) = 2A_2$ which is no greater than 2 since $A_2 = \pm 1$. If $C_2 = -1$, one finds $A_2B_2(C_2 - 1) + A_2(C_2 + 1) = -2A_2B_2$ which is not greater than 2 since $A_2, B_2 = \pm 1$.

8. $B_1 = C_1 = 1, A_1 = -1$. The inequality (4) can be written as $A_2B_2(C_2 - 1) + B_2(C_2 + 1) \le 2$. If $C_2 = 1$, one finds $A_2B_2(C_2 - 1) + B_2(C_2 + 1) = 2B_2$ which is no greater than 2 since $B_2 = \pm 1$. If $C_2 = -1$, one finds $A_2B_2(C_2 - 1) + B_2(C_2 + 1) = -2A_2B_2$ which is not greater than 2 since $A_2, B_2 = \pm 1$.

Thus, in each case, the inequality is satisfied regardless which values $A_i, B_j, C_k(i, j, k = 1, 2)$ take. The above inequality (4) do not include the terms of single correlation function, it is symmetric under the permutation of A_j and B_j. Moreover, by setting appropriate values of C_1 and C_2, the inequality reduces directly to the CHSH inequality for two-qubit. When $C_1 = -1, C_2 = 1$, the inequality is reduced to the CHSH inequality for two-qubit (up to a minus sign)

$$-Q(A_1B_1) - Q(A_1B_2) - Q(A_2B_1) + Q(A_2B_2) \le 2 \qquad (5)$$

3. Quantum Violation of the Bell Inequality for 3 Qubits

To test the quantum violation of any Bell inequalities, observables, we need to consider the quantum state. For the Bell type experiment in which three spatially separated observers Alice, Bob, and Charlie each measure two noncommuting observables $A_i = \hat{n}_{a_i} \cdot \vec{\sigma}(i = 1, 2)$ for Alice, $B_j = \hat{n}_{b_j} \cdot \vec{\sigma}(j = 1, 2)$ for Bob, and $C_k = \hat{n}_{c_k} \cdot \vec{\sigma}(k = 1, 2)$ for Charlie on the generalized GHZ states $|\psi\rangle$ of three qubits

$$|\psi\rangle = \cos\xi|0\rangle_A|0\rangle_B|0\rangle_C + \sin\xi|1\rangle_A|1\rangle_B|1\rangle_C \qquad (6)$$

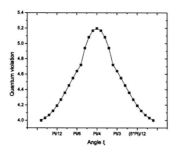

Figure 1. Numerical results for the generalized GHZ states $|\psi\rangle_{GHZ} = \cos\xi|000\rangle + \sin\xi|111\rangle$, which violate a Bell inequality for correlation functions (4) except $\xi = 0, \pi/2$. For the GHZ state with $\xi = \pi/4$, the quantum violation reaches its maximum value $3\sqrt{3} = 5.1965$.

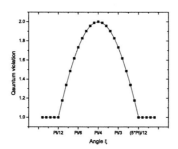

Figure 2. Numerical results for the generalized GHZ states $|\psi\rangle_{GHZ} = \cos\xi|000\rangle + \sin\xi|111\rangle$, which violate the Żukowski-Brukner inequality for 3 qubits except $(0, \pi/12), [7\pi/12, \pi/2)$. For the GHZ state with $\xi = \pi/4$, the quantum violation reaches its maximum value 4.

where $|k\rangle_i (k = 0, 1)$ describes kth basis state of the qubit $i(i = A, B, C)$ respectively. For each set of observables $A_i, B_j,$ and C_k,

$$A_i = \hat{n}_{a_i} \cdot \vec{\sigma} = \begin{pmatrix} \cos\theta_{a_i} & \sin\theta_{a_i} e^{-i\phi_{a_i}} \\ \sin\theta_{a_i} e^{i\phi_{a_i}} & -\cos\theta_{a_i} \end{pmatrix}$$

$$B_j = \hat{n}_{b_j} \cdot \vec{\sigma} = \begin{pmatrix} \cos\theta_{b_j} & \sin\theta_{b_j} e^{-i\phi_{b_j}} \\ \sin\theta_{b_j} e^{i\phi_{b_j}} & -\cos\theta_{b_j} \end{pmatrix}$$

$$C_k = \hat{n}_{c_k} \cdot \vec{\sigma} = \begin{pmatrix} \cos\theta_{c_k} & \sin\theta_{c_k} e^{-i\phi_{c_k}} \\ \sin\theta_{c_k} e^{i\phi_{c_k}} & -\cos\theta_{c_k} \end{pmatrix} \quad (7)$$

where $i, j, k = 1, 2$, the following correlation functions are resulted,

$$Q(A_i, B_j, C_k) = \langle \psi | A_i \otimes B_j \otimes C_k | \psi \rangle$$
$$Q(A_i, B_j) = \langle \psi | A_i \otimes B_j \otimes 1 | \psi \rangle$$
$$Q(B_j, C_k) = \langle \psi | 1 \otimes B_j \otimes C_k | \psi \rangle$$
$$Q(A_i, C_k) = \langle \psi | A_i \otimes 1 \otimes C_k | \psi \rangle \tag{8}$$

Numerical results show that the inequality (4) is violated by the generalized GHZ states for the whole region except $\xi = 0, \pi/2$, see Fig.1. When $\xi = \pi/4$, i.e., GHZ state given, the maximal quantum violation is $3\sqrt{3}$. For comparation, we show the numerical results for Żukowski-Brukner inequality in Fig.2. We also consider the strength of violation or visibility (V) [14] as the minimal amount V of the given entangled state $|\psi\rangle$ that one has to add to pure noise, ρ_{noise}, so that the resulting state violates local realism. The quantity V is thus the threshold visibility above which the state cannot be described by local realism, and it is sometimes called the critical visibility. Given the inequality (4), the admixed GHZ state cannot be described by local realism if and only if $V > 4\sqrt{3}/9$. For W state $|\psi\rangle_w = (|100\rangle + |010\rangle + |001\rangle)\sqrt{3}$, the critical visibility is $V_W = 0.7312$. These results are same as those given in Ref. [13]. Thus, the inequality (4) is an equivalent form of the one in [13]. Although inequality (4) is violated by any pure entangled states of three qubits as shown in [13], the visibility of GHZ state is not optimal. The visibility of the inequality for three qubits given by Żukouski-Brukner for GHZ state is 0.5. The improvement of this paper is that a Bell inequality involving corraltion functions, which is violated by the generalized GHZ states for the whole region, is constructed. However, there is no inequality which is not only maximally violated by GHZ state, but also violated by any pure entangled states. To develop such a new Bell inequality for three qubits is still an open problem.

4. Conclusion

In conclusion, we present a Bell inequality involving correlation functions for three qubit systems. The inequality is violated by generalized GHZ states of 3 qubits for the whole region, although it is not maximally violated by GHZ state. The visibility of GHZ state for the inequality is $4\sqrt{3}/9$, which is the same as that for the inequality given by us in Ref. [13]. The inequality (4) is the correlation function version of the one in [13].

This work is supported by NUS academic research Grant No. WBS: R-144-000-089-112. J.L. Chen acknowledges financial support from Singapore Millennium Foundation.

References

1. A. Einstein, B. Podolsky and N. Rosen, Phys. Rev. **47**, 777 (1935).
2. D. Bohm, *Quantum Physics* (Prentice Hall, 1951).
3. J. S. Bell, Physics 1, 195 (1964).
4. N. Gisin, Phys. Lett. A **154**, 201 (1991); N. Gisin and A. Peres, Phys. Lett. A **162**, 15-17 (1992).
5. M. Horodecki, P. Horodecki, and R. Horodecki, Phys. Rev. Lett. **84**, 2014 (2000).
6. K. G. H. Vollbrecht and R. F. Werner, J. Math. Phys. **41**, 6772 (2000).
7. M. Żukowski and Č. Brukner, Phys. Rev. Lett. **88**, 210401 (2002).
8. R. F. Werner and M. M. Wolf, Phys. Rev. A **64**, 032112 (2001).
9. V. Scarani and N. Gisin, J. Phys. A **34**, 6043 (2001).
10. D. M. Greenberger, M. Horne, A. Shimony, and A. Zeilinger, Am. J. Phys. **58**, 1131 (1990).
11. N. D. Mermin, Phys. Rev. Lett. **65**, 1838 (1990); M. Ardehali, Phys. Rev. A **46**, 5375 (1992); A. V. Belinskii and D. N. Klyshko, Phys. Usp. **36**, 653 (1993).
12. M. Żukowski, Č. Brukner, W. Laskowski, and M. Wiesniak, Phys. Rev. Lett. **88**, 210402 (2002).
13. J.L. Chen, C.F. Wu, L.C. Kwek and C.H. Oh, Phys. Rev. Lett. **93**, 140407 (2004).
14. D. Kaszlikowski, P.Ganciński, M. Żukowski, W. Mislaszewski and A. Zeilinger, Phys. Rev. Lett. **85**, 4418 (2000).

MINIMUM-ERROR DISCRIMINATION AMONG MIXED QUANTUM STATES

CHIH-LUNG CHOU

Department of Physics,
Chung Yuan Christian University,
Taoyuan 32023, Taiwan
E-mail: choucl@cycu.edu,tw

We review the problem of discriminating with minimum error among mixed quantum states, with emphasize on the analytically solved quantum ensembles.

1. Minimum-Error Discrimination

Quantum state discrimination is of fundamental importance for quantum communication and quantum cryptography. The theory of quantum communication is a well-developed field of research that concerns the transmission of information using quantum states and channels[1,2,3]. The transmission party encodes a message onto a set of quantum states $\{\rho_k\}$ with prior probability p_k for each of the states ρ_k. The set of signal states and the prior probabilities are also known to the receiving party. The task of the receiving party is to decode the received message, i.e., finding the best measurement strategy based upon the knowledge of the signal states and their prior probabilities. One possibility is to choose the strategy that minimizes the probability of detection error. In this paper we will review the problem of minimum-error discrimination with emphasize on the quantum ensembles of mixed quantum states.

In general, the measurement strategy is described in terms of a set of non-negative definite operators called the probability operator measure (POM) [1,2]. The measurement outcome labelled by "k" is associated with the element π_k of POM that has all the eigenvalues be either positive or zero, i.e, $\pi_k \geqslant 0$. The POM elements must sum into the identity operator $\sum_k \pi_k = \hat{1}$. The probability that the receiver will observe the outcome k given that the transmitted signal is ρ_j is $P(k|j) = \mathtt{tr}(\pi_k\rho_j)$. Here $\mathtt{tr}$ denotes the trace operation. It follows that the error probability is defined as

$$P_{error} = 1 - \sum_k p_k \operatorname{tr}(\pi_k \rho_k). \tag{1}$$

1.1. *Holevo's derivation for optimization conditions*

The necessary and sufficient conditions that lead to the minimum error probability are known to be [1,2,4,5]

$$\pi_k(p_k \rho_k - p_j \rho_j)\pi_j = 0, \tag{2}$$

$$\sum_k p_k \pi_k \rho_k - p_j \rho_j \geqslant 0. \tag{3}$$

The first condition holds for all j and k. The second condition means that all the eigenvalues of the operator at the left-hand side are non-negative and it holds for all j. These optimization conditions can be elegantly derived by using Holevo's derivation[2].

Let $\{U_{ik}\}$ be a set of N^2 operators that satisfy the conditions $\sum_i U_{im}^\dagger U_{ik} = \delta_{mk}\hat{1}$ for all $m,k = 1,2,\cdots,N$. Define a set of operators $s_i \equiv \sum_i U_{ik}\pi_k^{1/2}$ and a set of Hermitian operators $\pi_i' \equiv s_i^\dagger s_i$, $i = 1,2,\cdots,N$. It follows immediately that $\{\pi_i'\}$ also forms a new set of POM elements. Consider the infinitesimal transformation on $\{\pi_i^{1/2}\}$, $s_q = \sum_r U_{qr}\pi_r^{1/2}$, such that only the m-th and the k-th elements are transformed

$$s_q = \pi_q^{1/2}, \qquad\qquad q \neq k,m. \tag{4}$$

$$s_k = \pi_k^{1/2} - \varepsilon A^\dagger \pi_m^{1/2} \tag{5}$$

$$s_m = \pi_m^{1/2} + \varepsilon A \pi_k^{1/2}. \tag{6}$$

Here the real infinitesimal parameter $\varepsilon \ll 1$ is assumed and A denotes some arbitrary operator. To the order of ε^2 the new POM elements π_q' become

$$\pi_q' = \pi_q, \qquad\qquad q \neq k,m. \tag{7}$$

$$\pi_k' = \pi_k - \varepsilon(\pi_k^{1/2} A^\dagger \pi_m^{1/2} + \pi_m^{1/2} A \pi_k^{1/2}) \tag{8}$$

$$\pi_m' = \pi_m + \varepsilon(\pi_m^{1/2} A \pi_k^{1/2} + \pi_k^{1/2} A^\dagger \pi_m^{1/2}). \tag{9}$$

Under the infinitesimal transformation the error probability transforms as

$$1 - \sum_r p_r \operatorname{tr}(\pi_r' \rho_r) = 1 - \sum_r p_r \operatorname{tr}(\pi_r \rho_r)$$
$$+ 2\varepsilon \operatorname{Re}\{\operatorname{tr}[\pi_k^{1/2}(p_k \rho_k - p_m \rho_m)\pi_m^{1/2} A]\}. \tag{10}$$

Therefore if $\{\pi_r\}$ are the optimum POM elements then the following necessary conditions must hold

$$\pi_k^{1/2}(p_k\rho_k - p_m\rho_m)\pi_m^{1/2} = 0 \tag{11}$$

$$\Rightarrow \pi_k(p_k\rho_k - p_m\rho_m)\pi_m = 0, \text{ for all } m,k. \tag{12}$$

The sufficient conditions can be easily understood as follows. Suppose that $\{\pi_j\}$ is the best measurement strategy and $\{\pi'_j\}$ is a set of POM operators with the error probability P'. The difference between P' and the minimum error probability P_{min} must always be non-negative

$$P' - P_{min} = \sum_m \text{tr}[(\sum_k p_k\rho_k\pi_k - p_m\rho_m)\pi'_m] \geq 0. \tag{13}$$

It is then easy to see that the sufficient conditions in Eq. (3) satisfies Eq. (13).

The necessary and sufficient conditions in Eq. (2) and Eq. (3) are highly non-trivial and nonlinear in nature such that the required POM elements for the best measurement strategy are not easily derived from the conditions. In fact, only some classes of quantum ensembles are known for their best measurement strategies. These include the cases of only two signal states [1,6], symmetric states [5,7,8], mirror-symmetric states [9,10], linearly independent states [11], and equiprobable states that are complete in the sense that a weighted sum of projectors onto the states equals the identity operator [4]. Up to date there is no general way of obtaining analytical solutions for the problem of minimum-error detection. All the analytically solved quantum ensembles are solved in the case-by-case situations. In this review we discuss only the quantum ensembles of mixed quantum states. Instead of discussing the problem by numerical methods, we present three examples of mixed quantum states that are analytically solved in literature. First, we discuss the discrimination between two quantum states[1,6], either pure or mixed, in the next section. In the third section, we provide the optimum measurement strategy for discriminating aomng the mixed quantum states with symmetry[7,10]. We will show how to assume the possible forms of the best POM by symmetry considerations. In the last section, we make our conclusion.

2. Discrimination between Two Quantum States

Consider two quantum states ρ_1 and ρ_2 with *a priori* probabilities $p_1 = p$ and $p_2 = 1 - p$, respectively. The best measurement strategy for distinguishing between the two quantum states with minimum error requires the following necessary and sufficient conditions

$$\pi_1(p_1\rho_1 - p_2\rho_2)\pi_2 = 0 \tag{14}$$

$$p_1\pi_1\rho_1 + p_2\pi_2\rho_2 - p_1\rho_1 \geqslant 0 \tag{15}$$

$$p_1\pi_1\rho_1 + p_2\pi_2\rho_2 - p_2\rho_2 \geqslant 0, \tag{16}$$

where π_1 and π_2 are the operators for detecting ρ_1 and ρ_2, respectively. Now let $\bar{\rho} \equiv (p_1\rho_1 - p_2\rho_2)$ and rewrite Eq. (14) as $\pi_1\bar{\rho} = \pi_1\bar{\rho}\pi_1$. The necessary condition is easily satisfied if π_1 is a projection operator that satisfies $\pi_1^2 = \pi_1$ and commutes with $\bar{\rho}$. This means that both the detection operators π_1, π_2 as well as $\bar{\rho}$ can be written in terms of the same orthonormal basis $\{|\phi_a\rangle\}$.

$$\bar{\rho} = \sum_{a+}\eta_{a+}|\phi_{a+}\rangle\langle\phi_{a+}| + \sum_{a-}\eta_{a-}|\phi_{a-}\rangle\langle\phi_{a-}| \tag{17}$$

$$\pi_1 = \sum_{a+}|\phi_{a+}\rangle\langle\phi_{a+}| \tag{18}$$

$$\pi_2 = \sum_{a-}|\phi_{a-}\rangle\langle\phi_{a-}|, \tag{19}$$

where η_{a+} and η_{a-} are eigenvalues of $\bar{\rho}$. From the sufficient conditions in Eq. (15) and Eq. (16), the eigenvalues are found to satisfy $\eta_{a+} \geqslant 0$ and $\eta_{a-} \leqslant 0$. With the best POM, the minimum error probability is $P_{error} = 1 - tr(\sqrt{\bar{\rho}^\dagger\bar{\rho}})$.

It is interesting to note that the best strategy found in the case of two quantum states can actually be applied to either pure or mixed quantum states. Both the best detection operators are projection operators. One will project states onto the subspace spanned by the eigenvectors of $\bar{\rho}$ with non-negative eigenvalues. The other will project states onto the remaining subspace of $\bar{\rho}$.

3. Discrimination among Quantum States with Symmetry

Due to the success in finding analytical solutions for the problem of discriminating between two arbitrary quantum states, one may think that the analytical solutions to the problem of detecting three or more arbitrary quantum states can also be easily obtained. Unfortunately it is not true. Usually numerical methods are used in searching the best measurement strategies for quantum ensembles of more than two quantum states[12,13,14]. Analytical solutions are solved only in the cases with symmetric states or linearly independent states. It is understandable since if the given quantum ensemble is invariant under some symmetry G, then the minimum error probability for detecting the quantum states is also invariant under G. It implies that the best POM should respect the same symmetry G. This observation provides the clues in solving the problem of quantum state discrimination. In addition to the necessary and sufficient conditions in Eq. (2) and Eq. (3), the symmetry G imposes more restrictions on the possible forms of the best detection operators $\{\pi_k\}$. With the new restrictions imposed by the symmetry, the best measurement strategy may be found with ease.

3.1. *Discriminating among Symmetric Mixed Quantum States*

In this subsection, we will solve the optimum measurement strategy for a set of N mixed symmetric quantum states $\{\rho_k\}$ by symmetry consideration[7]. Assume that the quantum states are of equal prior probabilities $p_k = 1/N$ and respect the Z_N symmetry

$$\rho_k = R^k \rho_0 R^{\dagger k}, \qquad k = 0, 1, \cdots, (N-1), \tag{20}$$

$$R^N = \pm \hat{\mathbf{1}}, \tag{21}$$

where the operator R denotes the relevant part of the symmetry operator that lives in the same Hilbert subspace of the signal states $\{\rho_k\}$. $\hat{\mathbf{1}}$ denotes the identity operator of the Hilbert subspace of the signal states. We also assume R to be unitary $(RR^{\dagger} = R^{\dagger}R = \hat{\mathbf{1}})$ and nondegenerate, i.e., all its eigenvalues $\{b_\lambda\}$ are different for different eigenstates $\{|\lambda\rangle\}$. Therefore the dimensionality of R cannot be larger than the number of the signal states N otherwise at least two of the eigenvalues of R will be the same. Besides, we also assume that at least one of the the the signal states (assigned to be ρ_0) can be made to have all it matrix elements be real and non-negative, i.e., $\langle \lambda | \rho_0 | \lambda' \rangle \geqslant 0$ for some chosen set of the eigenstates $\{|\lambda\rangle\}$ of the operator R. It is noted that it is not always possible to satisfy this requirement.

In many of the cases [5,15] where the optimum strategies are known to be the square-root measurements with POM elements

$$\pi_k = \Phi^{-\frac{1}{2}} (p_k \rho_k) \Phi^{-\frac{1}{2}}, \tag{22}$$

$$\Phi \equiv \sum_k p_k \rho_k \tag{23}$$

where ρ_k denotes the k-th quantum signal states to be discriminated, and Φ is invariant under the transformation R. However the square-root measurement is not necessarily the best measurement strategy for the symmetric mixed quantum states discussed in this subsection. Instead, we assume the invariant operator Φ as

$$\Phi \equiv \sum_{k=0}^{N-1} R^k \Gamma_0 R^{\dagger k}, \tag{24}$$

where $\Gamma_0 \equiv |\varphi_0\rangle\langle\varphi_0|$ is the rank-1 operator that is formed by some normalized pure quantum state $|\varphi_0\rangle$. From Eq. (24), Φ is also Hermitian and non-negative definite, and commutes with R. This implies that both R and Φ can be expanded in terms of the same orthonormal basis $\{|\lambda\rangle\}$ as

$$\Phi = \sum_\lambda a_\lambda |\lambda\rangle\langle\lambda|, \tag{25}$$

$$R = \sum_\lambda b_\lambda |\lambda\rangle\langle\lambda|, \tag{26}$$

where $a_\lambda = N|\langle\lambda|\varphi_0\rangle|^2$ for all λ. In general, it is difficult to obtain the POM elements that satisfy the necessary and sufficient conditions in Eq. (2) and Eq. (3). However, we can obtain an analytical solution to these conditions for the symmetric mixed quantum states described in Eq. (20) and Eq. (21).

Proposition. Given the mixed symmetric quantum states as described in the Eq. (20) and Eq. (21), the optimum measurement strategy that minimizes the error probability P_{error} is given by

$$\pi_k \equiv R^k \Phi_2 \Gamma_0 \Phi_2 R^{\dagger k}, \qquad k = 0, 1, \cdots, (N-1). \tag{27}$$

$$\Gamma_0 \equiv |\varphi_0\rangle\langle\varphi_0|, \tag{28}$$

where $|\varphi_0\rangle$ is chosen such that $\langle\lambda|\varphi_0\rangle$ is real for all $|\lambda\rangle$ and satisfies $\langle\lambda|\varphi_0\rangle \neq 0$. The operator Φ_2 is defined by $\Phi_2 \equiv \sum_\lambda c_\lambda |\lambda\rangle\langle\lambda|$ with $c_\lambda \equiv N^{-\frac{1}{2}} \langle\lambda|\varphi_0\rangle^{-1}$.

It is noted that Φ_2 is Hermitian and also commutes with the symmetry operator R. The square of Φ_2 equals the inverse of Φ, i.e., $\Phi_2^2 = \Phi^{-1}$. The operator Φ_2 becomes the inverse square-root of Φ only when all $\langle\lambda|\varphi_0\rangle$ are real and positive. Obviously, the detection operators $\{\pi_k\}$ do respect the same symmetry R as the quantum states $\{\rho_k\}$ do. Similar to square-root measurement, the best POM elements proposed in the proposition are also rank-1 operators.

Proof of the Proposition. We need to prove that the POM elements defined in Eq. (27) and Eq. (28) are indeed POM elements and satisfy the necessary and sufficient conditions in Eq. (2) and Eq. (3). From Eq. (27) and Eq. (28), we can prove that all $\pi_k \geq 0$ as follows

$$\langle\phi|\pi_k|\phi\rangle = |\langle\phi|R^k\Phi_2|\varphi_0\rangle|^2 \geq 0, \qquad \text{for arbitrary } k, |\phi\rangle. \tag{29}$$

We can also *see* that $\pi_0 \geq 0$ by expanding π_0 in the basis $\{|\lambda\rangle\}$, $\pi_0 = \frac{1}{N}\sum_{\lambda,\lambda'} |\lambda\rangle\langle\lambda'|$. Under the basis all the matrix elements of π_0 equals $1/N$, thus π_0 has only one non-vanishing eigenvalue 1. The requirement that all eigenvalues of R are different guarantees that all the POM elements sum into identity operator

$$\sum_{k=0}^{N-1} \pi_k = \sum_k R^k \pi_0 R^{-k}$$

$$= \frac{1}{N} \sum_k \sum_{\lambda,\lambda'} (\frac{b_\lambda}{b_{\lambda'}})^k |\lambda\rangle\langle\lambda'|$$

$$= \sum_\lambda |\lambda\rangle\langle\lambda| = \hat{1}. \tag{30}$$

We proceed to prove that the POM given by Eq. (27) does satisfy the necessary and sufficient conditions listed in Eq. (2) and Eq. (3). By taking Eq. (27) into Eq. (2), we find

$$\pi_k (p_k\rho_k - p_j\rho_j)\pi_j$$

$$= \frac{1}{N} R^k \Phi_2 |\varphi_0\rangle\langle\varphi_0|\Phi_2(\rho_0 R^{j-k} - R^{j-k}\rho_0)\Phi_2|\varphi_0\rangle\langle\varphi_0|\Phi_2 R^{-j}. \tag{31}$$

By using Eq. (25) and Eq. (26) and considering the fact that all $\langle\lambda|\rho_0|\lambda'\rangle$ and $\langle\lambda|\varphi_0\rangle$ are real, we derive the following identity thus prove that Eq. (31) actually equals zero:

$$\langle \varphi_0|\Phi_2(\rho_0 R^{j-k} - R^{j-k}\rho_0)\Phi_2|\varphi_0\rangle$$

$$= \sum_{\lambda,\lambda'} c_\lambda c_{\lambda'} \langle\varphi_0|\lambda\rangle\langle\lambda|\rho_0|\lambda'\rangle\langle\lambda'|\varphi_0\rangle (b_{\lambda'}^{j-k} - b_\lambda^{j-k})$$

$$= 0. \tag{32}$$

The condition in Eq. (3) is proved as follows. First we observe that $\sum_k \pi_k\rho_k$ is Hermitian from

$$\sum_k \pi_k\rho_k = \sum_k R^k \pi_0\rho_0 R^{-k}$$

$$= \frac{1}{N} \sum_k \sum_{\lambda,\lambda',\lambda''} (\frac{b_\lambda}{b_{\lambda''}})^k |\lambda\rangle\langle\lambda'|\rho_0|\lambda''\rangle\langle\lambda''|$$

$$= \sum_{\lambda,\lambda'} |\lambda\rangle\langle\lambda'|\rho_0|\lambda\rangle\langle\lambda|$$

$$= \sum_k \rho_k\pi_k. \tag{33}$$

Therefore the operators $(\sum_k \pi_k\rho_k - \rho_j)$ are also Hermitian for all j. By sandwiching $(\sum_k \pi_k \rho_k - \rho_0)$ using an arbitrary state $|\phi\rangle$, we have

$$\langle\phi|\sum_k \pi_k \rho_k - \rho_0|\phi\rangle$$

$$= \sum_{\lambda,\lambda'}(|\langle\phi|\lambda\rangle|^2 - \langle\phi|\lambda'\rangle\langle\lambda|\phi\rangle)\langle\lambda'|\rho_0|\lambda\rangle$$

$$= \frac{1}{2}\sum_{\lambda,\lambda'}(|\langle\phi|\lambda\rangle|^2 + |\langle\phi|\lambda'\rangle|^2 - \langle\phi|\lambda'\rangle\langle\lambda|\phi\rangle - \langle\phi|\lambda\rangle\langle\lambda'|\phi\rangle)\langle\lambda'|\rho_0|\lambda\rangle$$

$$= \frac{1}{2}\sum_{\lambda,\lambda'}\varepsilon_{\lambda\lambda'}(\phi)\langle\lambda'|\rho_0|\lambda\rangle \geqslant 0, \tag{34}$$

$$\varepsilon_{\lambda\lambda'}(\phi) \equiv ((\langle\lambda|\phi\rangle - \langle\lambda'|\phi\rangle)(\langle\lambda|\phi\rangle - \langle\lambda'|\phi\rangle))^\dagger \geqslant 0. \tag{35}$$

From Eq. (34) we conclude that $(\sum_k \pi_k \rho_k - \rho_0)$ is a Hermitian operator and non-negative definite. This then leads to the fact that $(\sum_k \pi_k \rho_k - \rho_j)$ are also non-negative definite and Hermitian for all possible j since $(\sum_k \pi_k \rho_k - \rho_j) = R^j(\sum_k \pi_k \rho_k - \rho_0)R^{\dagger j}$.

In this case, the best measurement strategy is not necessarily the square-root measurement in literature[2,5,15]. In fact it is shown that the square-root measurement can be a special case when the discriminated symmetric quantum states are pure quantum states[7]. It also shows that the optimum POM elements for detecting the symmetric mixed quantum states are all rank-1 operators, independent of the rank of the discriminated quantum states. The solution demonstrates that the rank of the best POM elements could be much smaller (rank=1 in the case) than the rank of the signal states.

3.2. Discriminating among Mirror-Symmetric Mixed Quantum States

In this subsection we discuss the problem of discriminating among three mixed quantum states that respect the mirror symmetry T

$$T : \begin{cases} |+\rangle \longrightarrow |+\rangle, \\ |-\rangle \longrightarrow -|-\rangle, \end{cases} \tag{36}$$

where $|+\rangle = (1,0)$ and $|-\rangle = (0,1)$ are orthonormal basis states in the two-dimensional Hilbert space. The three quantum states $\{\rho_0, \rho_1, \rho_2\}$ are defined as

$$\rho_0 = a_0\sigma_0 + a_3\sigma_3, \tag{37}$$

$$\rho_1 = a_0\sigma_0 + a_1\sigma_1 + a_2\sigma_2 + a_4\sigma_3, \tag{38}$$

$$\rho_2 = a_0\sigma_0 - a_1\sigma_1 - a_2\sigma_2 + a_4\sigma_3, \tag{39}$$

with prior probabilities p_0, p_1 and p_2, respectively. Here $(\sigma_1, \sigma_2, \sigma_3)$ denote the Pauli matrices

$$\sigma_1 = \begin{pmatrix} 0 & 1 \\ 1 & 0 \end{pmatrix}, \sigma_2 = \begin{pmatrix} 0 & -i \\ i & 0 \end{pmatrix}, \sigma_3 = \begin{pmatrix} 1 & 0 \\ 0 & -1 \end{pmatrix}, \tag{40}$$

and σ_0 is the identity matrix in the two-dimensional Hilbert space. The coefficients a_k are all real, and satisfy $a_0 = 1/2$ and $a_0 \geqslant \sqrt{a_1^2 + a_2^2 + a_4^2}$ as implied by the condition that all the quantum signals ρ_k are non-negative definite and Hermitian with $\text{tr}(\rho_k) = 1$. The coefficients a_1 and a_3 are assumed to satisfy $a_1 \geqslant 0$ and $-1/2 \leqslant a_3 \leqslant 0$ without loss of generality. By mirror symmetry, two of the three prior probabilities are equal $p_1 = p_2 = p$. It then leads to $p_0 = 1 - 2p$ and $0 \leqslant p \leqslant 1/2$. It is easy to see that $\{\rho_0, \rho_1, \rho_2\}$ transform as $(\rho_0 \rightarrow \rho_0, \rho_1 \leftrightarrow \rho_2)$ under the mirror symmetry, thus form an invariant set of the mirror symmetry.

On the other hand, we expect that the optimum POM elements also respect the mirror symmetry, i.e., $(\pi_0 \rightarrow \pi_0, \pi_1 \leftrightarrow \pi_2)$. It thus restricts the possible forms of the optimum POM elements. The optimum POM elements can be parameterized as follows:

$$\begin{aligned} \pi_1 &= b_0\sigma_0 + b_1\sigma_1 + b_2\sigma_2 + b_3\sigma_3, \\ \pi_2 &= b_0\sigma_0 - b_1\sigma_1 - b_2\sigma_2 + b_3\sigma_3, \\ \pi_0 &= (1 - 2b_0)\sigma_0 - 2b_3\sigma_3, \end{aligned} \tag{41}$$

where all the parameters b_k are real numbers. With the POM elements in Eq. (41), the error probability in Eq. (1) is also invariant under the mirror symmetry.

3.2.1. *The optimum measurement strategy*

Our goal is to minimize the error probability P_{error} in Eq. (1). The minimization problem is indeed a linear programming problem since the error probability P_{error} is linear in all the parameters b_0, b_1, b_2 and b_3. However, the parameters are not arbitrary. They must satisfy the condition that the measurement strategy is a set of non-negative definite and Hermitian operators, i.e, $\pi_k \geqslant 0$ for all k. This leads to the following constraints

$$b_0 \geqslant \sqrt{b_1^2 + b_2^2 + b_3^2}, \tag{42}$$

$$1 \geqslant 2b_0 + 2|b_3|. \tag{43}$$

The constraints indicate that the allowed parameters (b_0, b_1, b_2, b_3) form a cone-like region $(1/2 \geqslant b_0 \geqslant \sqrt{b_1^2 + b_2^2 + b_3^2})$ bounded by two planes $(1 \geqslant 2b_0 + 2b_3, 1 \geqslant 2b_0 - 2b_3)$ in the parameter space. Considering that P_{error} is linear in

b_0, b_1, b_2 and b_3, P_{error} shall have the minimum at the surface of the allowed parameter region. In the remaining parts of the section, we will discuss the error probability function on various locations of the surface of the allowed parameter region and find the minimum of the error probability.

3.2.2. Two-POM-elements scenario

One of the interesting area on the surface of the allowed parameter region locates at the intersection of the cone surface $(1/2 = b_0 = \sqrt{b_1^2 + b_2^2 + b_3^2})$ and the two planes $(1 = 2b_0 + 2b_3, 1 = 2b_0 - 2b_3)$. It corresponds to the parameters $b_0 = 1/2$ and $b_3 = 0$. The other parameters (b_1, b_2) then satisfy $\sqrt{b_1^2 + b_2^2} = 1/2$. With these parameters, the POM elements become $\pi_0 = 0$, $\pi_1 = 1/2\sigma_0 + b_1\sigma_1 + b_2\sigma_2$ and $\pi_2 = 1/2\sigma_0 - b_1\sigma_1 - b_2\sigma$. From the optimization conditions in Eq. (2), we find that the parameters b_1 and b_2 must satisfy $b_1 = a_1/(2\sqrt{a_1^2 + a_2^2})$ and $b_2 = a_2 b_1/a_1$, respectively. The conditions in Eq. (3) then give a restriction on the prior probability p. We find that p must be no less than the critical prior probability P_{c2}, $p \geqslant P_{c2}$. The critical prior probability P_{c2} is totally determined by the coefficients a_k and defined as

$$
P_{c2} \equiv \frac{1 - 2a_3}{3 + 2\sqrt{a_1^2 + a_2^2} - 4a_3 - 2a_4}, \quad (-a_3) \geqslant \frac{-a_4}{1 + 2\sqrt{a_1^2 + a_2^2}}
$$

$$
\equiv \frac{1 + 2a_3}{3 + 2\sqrt{a_1^2 + a_2^2} + 4a_3 + 2a_4}, \quad (-a_3) \leqslant \frac{-a_4}{1 + 2\sqrt{a_1^2 + a_2^2}}. \tag{44}
$$

Therefore, the optimum POM has only two nonzero elements if $p \geqslant P_{c2}$. In this scenario, the optimum error probability is $P_{error} = 1 - p(1 + 2\sqrt{a_1^2 + a_2^2})$.

It is noted that both the nonzero optimum POM elements are rank-1 operators in this scenario. This conclusion is similar to the case that has three pure signal states[9]. In fact, our mixed signal states reduce to pure states at $a_3 = -1/2$ and $\sqrt{a_1^2 + a_2^2 + a_4^2} = 1/2$. When $a_1 = \sin(2\theta)/2$, $a_2 = 0$, $a_3 = -1/2$ and $a_4 = -\cos(2\theta)/2$, we reproduce the results as obtained in the paper by Andersson et al.[9].

3.2.3. One-POM-element scenario

Let us consider the situation when the prior probabilities p are small. If p is small enough then the mirror-singlet signal ρ_0 appears far more frequent than the other two signals ρ_1 and ρ_2. Therefore, the best measurement strategy may have only a single nonzero POM element $\pi_0 = \sigma_0$. The assumed strategy corresponds to the parameters $b_0 = b_1 = b_2 = b_3 = 0$, the cusp of the allowed region in the (b_0, b_1, b_2, b_3) parameter space. Since the parameters locate at the surface of the allowed parameter region, the assumed one-POM-element strategy may be optimum.

To support our assumption, we must prove that $(\pi_0 = \sigma_0, \pi_1 = \pi_2 = 0)$ satisfy the necessary and sufficient conditions in Eq. (2) and Eq. (3). It is easy to see that the strategy $(\pi_0 = \sigma_0, \pi_1 = \pi_2 = 0)$ is a trivial solution to the condition Eq. (2). To satisfy the condition Eq. (3), the prior probability p must be no greater than the critical prior probability P_{c1}:

$$p \leqslant \frac{3 - 8a_3^2 - 4a_3a_4 - 2\sqrt{(1 - 4a_3^2)(a_1^2 + a_2^2) + (a_3 - a_4)^2}}{9 - 4(a_1^2 + a_2^2) - (4a_3 + 2a_4)^2} \equiv P_{c1}. \qquad (45)$$

Therefore, the assumed one-POM-element strategy is optimum with the optimum error probability $P_{error} = 2p$ if $p \leqslant P_{c1}$.

It is noted that the one-POM-element scenario does not appear in the case that has three pure signal states as discussed in [9]. The three pure states correspond to the parameters $a_1 = \sin(2\theta)/2, a_2 = 0, a_3 = -1/2$, and $a_4 = -\cos(2\theta)/2$ in the section with vanishing critical prior probability $P_{c1} = 0$. This means that no one-POM-element scenario will exist in the case of three pure mirror-symmetric states[9] unless for $p = 0$.

3.2.4. *Three-POM-elements scenario*

The other surface area of the allowed parameter region that may give the optimum measurement strategy is the intersection of the two surfaces $(b_0 = \sqrt{b_1^2 + b_2^2 + b_3^2})$ and $(1 - 2b_0 = 2b_3)$, or the intersection of $(b_0 = \sqrt{b_1^2 + b_2^2 + b_3^2})$ and $(1 - 2b_0 = -2b_3)$. It is easily shown that the later intersection will not simultaneously satisfy the necessary and sufficient conditions in Eq. (2) and Eq. (3). The first intersection inevitably implies that all POM elements are rank-1 operators. We can re-parameterize the POM elements as

$$\pi_0 = \begin{pmatrix} 0 & 0 \\ 0 & 1 - x^2 \end{pmatrix}, \qquad \pi_1 = \frac{1}{2}\begin{pmatrix} 1 & xe^{-i\alpha} \\ xe^{i\alpha} & x^2 \end{pmatrix},$$

$$\pi_2 = \frac{1}{2}\begin{pmatrix} 1 & -xe^{-i\alpha} \\ -xe^{i\alpha} & x^2 \end{pmatrix}. \qquad (46)$$

The conditions in Eq. (2) then lead to

$$x = \frac{2p\sqrt{a_1^2 + a_2^2}}{1 - 3p - (2 - 4p)a_3 + 2pa_4}, \tag{47}$$

$$\alpha = \arctan(a_2/a_1). \tag{48}$$

The other optimization conditions in Eq. (3) will be satisfied automatically if p is in the range $P_{c1} \leqslant p \leqslant P_{c2}$. This result coincides with both the one-POM- and the two-POM-elements scenarios. In the three-POM-elements scenario, the success probability is $P_{success} \equiv (1 - P_{error})$

$$P_{success} = \{1 - 4p + 3p^2 + 4p^2(a_1^2 + a_2^2 + a_4^2) + 4pa_4(1 - 2a_3)(1 - 2p)$$
$$- 4a_3(1 - a_3)(1 - 2p)^2\}/\{2 - 6p - (4 - 8p)a_3 + 4pa_4\}. \tag{49}$$

4. Conclusion

In this paper we reviewed the problem of discriminating with minimum error among mixed quantum states, with emphasize on the analytically solved quantum ensembles. We discussed three quantum ensembles in the paper: quantum ensemble with two quantum states, N symmetric mixed quantum states, and three mixed quantum states with mirror-symmetry. In the case of distinguishing between two quantum states, the best POM elements are projection operators. One element will project states onto the space that is spanned by the normalized eigenvectors with non-negative eigenvalues of the operator $\bar{\rho} \equiv p_1\pi_1 - p_2\pi_2$. The other POM element projects states onto the space spanned by the remaining normalized eigenvectors of $\bar{\rho}$. This result is exact and easy to understand. However, it is not easily generalized to the cases of more than two quantum states, neither pure nor mixed.

The other two cases presented in this review are mixed quantum states with symmetry. Like the quantum ensembles of pure quantum states discussed in literature[5,9,8,11], mixed quantum states with symmetry could be easier to be solved analytically. In general we may expect that both the quantum ensembles $\{\rho_k\}$ and the best POM $\{\pi_k\}$ will respect the same symmetry. This expectation is reasonable since the error probability will remain unchanged under the symmetry if the detection is optimum. The possible forms of the detection operators π_k are then further restricted by by the symmetry. Hopefully we can use the new restrictions as well as the necessary and sufficient conditions in Eq.(2) and Eq. (3) for obtaining analytical solutions. Both the quantum ensembles discussed in Sec. 3 are solved analytically by the symmetry consideration. It is found that both of the best measurement strategies do respect the same symmetries as the corresponding quantum states do.

References

1. C. W. Helstrom, *Quantum Detection and Estimation Theory*, Academic Press, New York, 1976.
2. A. S. Holevo, *Probabilistic and Statiscal Aspects of Quantum Theory*, North-Holland, Amsterdam, 1982.
3. M. Sasaki, S. M. Barnett, R. Jozsa, M. Osaki, and O. Hirota, *Phys. Rev. A* **59**, 3325 (1999).; E. B. Davies, *IEEE Trans. Inf. Theory* **IT-24**, 596 (1978).
4. H. P. Yuen, R. S. Kennedy and M. Lax, *IEEE Trans. Inform. Theory* **IT-21**, 125 (1975).
5. M. Ban, K. Kurokawa, R. Momose and O. Hirota, *Int. J. Theor. Phys.* **36**, 1269 (1997);
6. U. Herzog, *J. Opt. B: Quantum Semiclassical Opt.* **6**, 24 (2004).
7. C.-L. Chou and L.Y. Hsu, *Phys. Rev. A* **68**, 042305 (2003).
8. S. M. Barnett, *Phys. Rev. A* **64**, 030303 (2001).
9. E. Andersson, S. M. Barnett, C. R. Gilson and K. Hunter, *Phys. Rev. A* **65**, 052308 (2002).
10. C.-L. Chou, *Phys. Rev. A* **70**, 062316 (2004).
11. U. Herzog and J. A. Bergou, *Phys. Rev. A* **65**, 050305 (2002).
12. Y. C. Eldar, A. Megretski, and G. C. Verghese, *IEEE Trans. Inform. Theory* **IT-49**, 1007 (2003).
13. M. Jezek, J. Rehacek, and J. Fiurasek, *Phys. Rev. A* **65**, 060301 (2002).
14. E. M. Rains,*IEEE Trans. Inform. Theory* **IT-47**, 2921 (2001); A. C. Doherty, P. A. Parrilo, and F. M. Spedalier, *Phys. Rev. Lett.* **84**, 187904 (2002).
15. Y. C. Eldar and G. D. Forney, *IEEE Trans. Inform. Theory* **IT-47**, 858 (2001); P. Hausladen and W. K. Wootters, *J. Mod. Opt.* **41**, 2385 (1994).

SECURITY OF QUANTUM KEY DISTRIBUTION WITH STRONG
PHASE-REFERENCE PULSE

MASATO KOASHI

Division of Materials Physics, Graduate School of Engineering Science,
Osaka University, 1-3 Machikaneyama, Toyonaka, Osaka 560-8531, Japan;
CREST Photonic Quantum Information Project,
4-1-8 Honmachi, Kawaguchi, 331-0012, Japan
E-mail: koashi@mp.es.osaka-u.ac.jp

In the BB84 protocol with a perfect single photon source, the key rate decreases linearly with the transmission η of the channel. If we simply replace this source with a weak coherent-state pulse, the key rate drops more rapidly (as $O(\eta^2)$) since the presence of multiple photons favors the eavesdropper. Here we discuss the unconditional security of a quantum key distribution protocol in which bit values are encoded in the phase of a weak coherent-state pulse relative to a strong reference pulse, which is essentially the one proposed by Bennett in 1992 (the B92 scheme). We show that in the limit of high loss in the transmission channel, we can construct a secret key with a rate proportional to the transmission η of the channel.

1. Introduction

Quantum key distribution (QKD) enables us to distribute a secret key between two distant parties, Alice and Bob, even if the quantum channel between them suffers from small noises. As long as the law of quantum mechanics is valid, an eavesdropper, Eve, cannot force Alice and Bob to accept a key on which she has a nonnegligible amount of information. A proof of such unconditional security was first provided by Mayers[1] for the BB84 protocol [2], followed by other proofs[3,4,5,6,7,8,9,10,11]. While a perfect single-photon source is assumed in the earlier proofs, recent proofs[6,7] cover the use of a weak laser pulse in a coherent state as a substitute for a single photon. This is good news in the practical point of view, but comes with a price: the multiphoton components of the weak pulse allow Eve a so-called photon-number splitting attack[12,13]. In order to achieve the security under this attack, Alice must lower the amplitude of her weak pulse as the loss in the channel increases. As a result, there is a bound [12] on the achievable key rate which scales as $O(\eta^2)$ with channel transmission η.

In this paper, we describe a proof of the unconditional security of a scheme using a weak coherent pulse and achieving a rate that scales as $O(\eta)$. The scheme is essentially the one proposed by Bennett[14], in which a strong pulse is transmitted as a phase reference together with a weak pulse containing the bit information in the relative phase. We made a minor modification to introduce a second local oscillator (LO) for Bob. This makes the analysis simpler, and allows us to assume a realistic threshold detector that may be noisy, inefficient, sensitive to multimodes of light, and only discriminates the vacuum from one or more photons. We show that in the limit of $\eta \rightarrow 0$, the rate of the key is proportional to η, and the limiting value of G/η is determined by the observed error rate and the counting rate of the detector.

This paper is organized as follows. In Sec. 2, we describe a QKD scheme with a strong phase-reference pulse. The proof of the security of this scheme against any intervention by an eavesdropper that is allowed by law of quantum mechanics will be given in Sec. 3, followed by an expression of the key rate in Sec. 4. The key rate in the limit of high loss is discussed in Sec. 5, and Sec. 6 concludes the paper.

2. QKD scheme with strong reference pulse

In this section, we introduce a QKD protocol, which is a modified version of the original B92 proposed by Bennett[14]. The sender encodes the bit value into the phase of a weak coherent-state pulse, and transmits it together with a strong phase-reference pulse. The only difference from the original B92 is the use of an additional local oscillator on the receiver's side.

The scheme is depicted in Fig. 1(a). Suppose that Alice's LO emits a strong pulse in a coherent state with complex amplitude $|\alpha_0|e^{i\phi_A}$. Using an asymmetric beamsplitter (BS1), Alice extracts a weak pulse with very small amplitude $\alpha = |\alpha|e^{i\phi_A}$, and encodes a randomly chosen bit value 0 or 1 by applying phase shift 0 or π, resulting in state $|\alpha\rangle$ or $|-\alpha\rangle$, respectively. Together with this signal, she sends the strong pulse from the other output of BS1 to Bob as a phase reference.

On the receiver's side, Bob chooses randomly a bit value 0 or 1, and applies phase shift 0 or π to the weak signal pulse, respectively. Instead of using the reference pulse from Alice directly, Bob uses another LO and tries to lock its phase to Alice's one. Suppose that Bob's LO produces a strong pulse with complex amplitude $|\beta_0|e^{i\phi_B}$. Combining a portion of this pulse and the reference pulse from Alice, he conducts a series of interference experiments (M) to infer the phase difference $\phi_A - \phi_B$. He then applies a phase shift equal to this estimated value ϕ^* to his LO, and mixes it with the weak signal from Alice at BS2. The mixed signal is measured by a threshold detector, which gives a "click" whenever it receives one or more photons. Bob reports the outcome of the detector to Alice over an authenticated public channel. The click implies a conclusive result, and both parties accept their bits. No click implies an inconclusive result, and they discard the bits.

The security analysis in this paper is valid even if LOs with phases ϕ_A and ϕ_B are available to Eve. Then, the reference pulse from Alice gives no information to Eve. The only effect of Eve's attack on this pulse is to disturb the measurement outcome ϕ^* to be deviated from the desired value, as $\phi^* = \phi_A - \phi_B - \Delta\phi$. But exactly the same effect can be obtained by just applying the phase shift $\Delta\phi$ to the weak signal from Alice (Eve may simulate M by herself). Hence we can safely assume that Eve simply ignores the strong reference pulse. Similarly, any imperfection in the estimation process M, including the fundamental limitation arising from finiteness of the amplitudes of the two LOs, has the same effect as introducing a noise source applying a phase shift $\Delta\phi$ on the weak signal while assuming a perfect estimation, $\phi^* = \phi_A - \phi_B$.

The major imperfections in the detector can be treated as follows. Suppose that the quantum efficiency of the detector is η_D, the transmission coefficient of BS2 is η_{BS2}, and the amplitude of LO incident on BS2 is $(1 - \eta_{BS2})^{-1/2}\eta_D^{-1/2}\beta$. Then, the same measurement can be implemented by inserting a lossy medium (BS3)

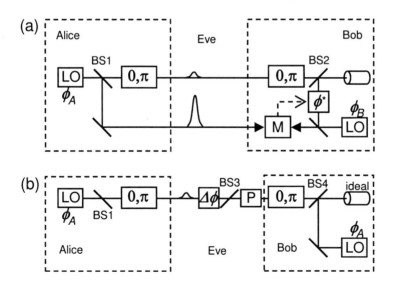

Figure 1. (a) A scheme with a strong reference pulse. (b) An equivalent scheme except that Eve's region is extended.

with transmission $\eta_{BS2}\eta_D\eta_0^{-1}$, then mixing LO with amplitude $(1-\eta_0)^{-1/2}\beta$ by a beamsplitter BS4 with transmission η_0, followed by a detector with unit efficiency. Here we take the limit of $\eta_0 \to 1$. The dark counting of the detector or the detection of stray photons can be simulated by a device (P) that inserts a photon in a mode that is orthogonal to the modes of the LOs. We thus finally arrive at a scheme with an ideal threshold detector and a locked pair of LOs, as in Fig. 1(b). In this figure, we have taken a conservative assumption to extend the region accessible by Eve for the sake of simplicity. If a protocol is secure with this scheme, the same protocol implemented by the scheme in Fig. 1(a) is also secure.

Bob's decision process in the scheme in Fig. 1(b) can be regarded as a generalized measurement on the light entering his site with three outcomes, 0, 1, and 2, where the last one means "inconclusive". Let $H_B = H_0 \otimes H_1 \otimes \cdots \otimes H_\nu \otimes \cdots$ be the Hilbert space for the light modes received by Bob that are sensible by the detector. The mode $\nu = 0$ represents the pulse mode of Bob's LO, and the modes with $\nu \geq 1$ are orthogonal to it. Let us write the coherent state $|\beta\rangle_0|0\rangle_1|0\rangle_2\cdots$ simply as $|\beta\rangle$. Then, the generalized measurement is described by the POVM $\{F_0, F_1, F_2\}$, where

$$F_0 = (\mathbf{1} - |-\beta\rangle\langle-\beta|)/2,$$
$$F_1 = (\mathbf{1} - |\beta\rangle\langle\beta|)/2,$$
$$F_2 = \mathbf{1} - F_0 - F_1. \tag{1}$$

If everything is ideal except for the transmission η in the channel, Alice's signal is received by Bob in coherent states $|\pm\sqrt{\eta}\alpha\rangle$, and they can agree on a key without errors by choosing $\beta = \sqrt{\eta}\alpha$.

3. Security proof

In this section, we describe a proof of the unconditional security of the scheme depicted in Fig. 1(b). As stated in the last section, the proof also establishes the same degree of security for the one in Fig. 1(a). The precise meaning of the unconditional security is the following. We allow an eavesdropper to make any attempt that is allowed by the law of quantum mechanics, including the coherent attack which uses a big auxiliary system to be interacted with every pulse transmitted from Alice. As usual, we assume that Alice and Bob can communicate through an authenticated classical channel. In practice, such a channel may only be established by consuming a small length of secret key which has been shared prior to the protocol. The successful protocol should produce more secret keys than is consumed for the authentication of the classical channel. In this sense, the function of the QKD protocol here is to amplify the amount of shared secret key, rather than to produce one from the scratch.

Before describing the proof of unconditional security, we introduce several notations. We decompose H_B as $H_B = K_B \oplus H_{ex}$, where K_B is the two-dimensional subspace spanned by $|\beta\rangle$ and $|-\beta\rangle$. We assume α and β to be real and positive without loss of generality. Let $\{|\mu_l\rangle_B\}_{l=1,2,...}$ be an arbitrary complete orthonormal basis for H_{ex}. We identify K_B as a qubit, and define its X basis as $\{|0_x\rangle_B \equiv (|\beta\rangle + |-\beta\rangle)/(2c_\beta), |1_x\rangle_B \equiv (|\beta\rangle - |-\beta\rangle)/(2s_\beta)\}$, where $2c_\beta^2 - 1 \equiv 1 - 2s_\beta^2 \equiv \langle -\beta|\beta\rangle = e^{-2|\beta|^2}$. The Z-basis states are denoted as $|j_z\rangle_B \equiv (|0_x\rangle_B + (-1)^j|1_x\rangle_B)/\sqrt{2}$ ($j = 0,1$). For Alice's side, we denote by H_A the Hilbert space of the light modes emitted from her site. We also introduce an auxiliary qubit in Alice's site, with Hilbert space K_A. We denote the X- and the Z-basis states as $|j_x\rangle_A$ and $|j_z\rangle_A$ ($j = 0,1$). We sometimes denote the projection $|\Phi\rangle\langle\Phi|$ as $P(|\Phi\rangle)$.

The key idea in the security proof is a trace-nonincreasing completely positive map, which is specified by Kraus operators $A_j : H_B \rightarrow K_B$ ($j = 0,1,2,...$) defined by

$$A_0 = s_\beta|0_x\rangle_B\langle 0_x| + c_\beta|1_x\rangle_B\langle 1_x|$$

for $j = 0$ and

$$A_j = |0_x\rangle_B\langle\mu_j|$$

otherwise. Since $\sum_j A_j^\dagger A_j \leq 1$, there exists a filter with the following property. It takes any state ρ acting on H_B as an input, and it accepts with probability $p = \sum_j \text{Tr}(A_j^\dagger A_j\rho)$ while it rejects with probability $1 - p$. Whenever it accepts, it returns the output state $\sum_j A_j\rho A_j^\dagger/p$ acting on K_B. This filter is related to the POVM $\{F_0, F_1, F_2\}$ by

$$F_k = \sum_j A_j^\dagger|k_z\rangle_B\langle k_z|A_j \tag{2}$$

for $k = 0,1$, which is easily confirmed. This relation implies that we can implement the measurement $\{F_0, F_1, F_2\}$ by applying the filter and conducting Z-basis measurement on the output state when it accepts (if it rejects, we assume that the outcome is "2").

With the above decomposition of Bob's measurement, we can prove the unconditional security by a method similar to the cases of qubit-based B92 protocols[9,10]. We introduce a protocol based on entanglement distillation[15], which is later shown to be equivalent to the real protocol. In the new protocol, (1) Alice prepares state $(|0_z\rangle_A|\alpha\rangle + |1_z\rangle_A|-\alpha\rangle)/\sqrt{2}$ on $K_A \otimes H_A$. We assume that Alice produces $2N$ copies of this state. (2) Eve receives $2N$ pulses (corresponding to $H_A^{\otimes 2N}$) from Alice, and prepares a state on $H_B^{\otimes 2N}$, which may be entangled to Eve's system. (3) After Bob has received $2N$ pulses (corresponding to $H_B^{\otimes 2N}$), Alice and

Bob randomly permutate the order of $2N$ pairs of systems by public discussion. (4) For the first N pairs (check pairs), Alice measures each qubit (K_A) on Z basis, and Bob performs the POVM $\{F_0, F_1, F_2\}$ on each pulse (H_B). They disclose all the results, and learn the number n_{err} of error events where the combination of Alice's and Bob's outcomes are $(0, 1)$ or $(1, 0)$. (5) For the other N pairs (data pairs), Bob applies the above filter to each pulse, and discloses each result (accept or reject). Let n_{fil} be the number of events where the filter has accepted. (6) Alice and Bob now have n_{fil} pairs of qubits ($K_A \otimes K_B$), from which they try to extract a number of pairs in the maximally entangled state $(|0_z\rangle_A|0_z\rangle_B + |1_z\rangle_A|1_z\rangle_B)/\sqrt{2}$. To do so, they estimate the number n_{bit} of pairs with a bit error (represented by the subspace spanned by $\{|0_z\rangle_A|1_z\rangle_B, |1_z\rangle_A|0_z\rangle_B\}$) and the number n_{ph} of pairs with a phase error (the subspace spanned by $\{|0_x\rangle_A|1_x\rangle_B, |1_x\rangle_A|0_x\rangle_B\}$), from the knowledge of n_{fil} and n_{err}. If neither number of errors is too high, they run an entanglement distillation protocol (EDP) and then measure on Z basis to determine the final key. As in the proof of BB84[5], if the estimation of the upper bounds for n_{bit} and n_{ph} is correct except for a probability that becomes exponentially small as N increases, this protocol is essentially secure.

According to the argument by Shor and Preskill[5], if we choose an appropriate EDP scheme, Alice and Bob can conduct Z-basis measurement on the n_{fil} pairs immediately after step (5) and decide the final key by a public discussion without compromising the security. Then, Eq. (2) shows that Bob's measurement on each data qubit is also the POVM $\{F_0, F_1, F_2\}$. Alice's measurement can be further brought forward to the end of step (1), then this step is equivalent to just preparing state $|\alpha\rangle$ or $|-\alpha\rangle$ randomly. The new protocol is thus equivalent to the prepare-measure protocol implemented as in Fig. 1(b).

The remaining task for the security proof is to establish an upper bound n_{bitmax} on n_{bit}, and n_{phmax} on n_{ph}, such that the probability of exceeding these bounds is exponentially small. Since n_{err} and n_{bit} are the results of the same measurement applied to the (randomly assigned) check pairs and to the data pairs, we can apply a classical probability estimate to see that $|n_{bit} - n_{err}| \leq N\varepsilon$ holds except for a small probability which is asymptotically smaller than $\sim \exp(-N\varepsilon^2)$. Hence we take

$$n_{bitmax} = n_{err} + N\varepsilon \tag{3}$$

The estimation of n_{ph} can be done by considering what could have happened if Alice and Bob measured their n_{fil} pairs of data qubits in X basis and determined n_{ph} by discussion, just after the step (5). In this scenario, they obtain three numbers $(n_{fil}, n_{ph}, n_{err})$. The following argument shows that some combinations of $(n_{fil}, n_{ph}, n_{err})$ are exponentially rare for any attack by Eve, and hence gives an (exponentially reliable) upper bound $n_{\text{phmax}}(n_{\text{fil}}, n_{\text{err}})$ for n_{ph} as a function of the other two.

We can regard n_{ph} as the number of events where a measurement on $K_A \otimes H_B$ produced the outcome corresponding to the element of a POVM

$$M_{\text{ph}} \equiv \sum_j P(|0_x\rangle_A) \otimes A_j^\dagger |1_x\rangle_B \langle 1_x | A_j + P(|1_x\rangle_A) \otimes A_j^\dagger |0_x\rangle_B \langle 0_x | A_j$$
$$= s_\beta^2 P(|1_x\rangle_A |0_x\rangle_B) + c_\beta^2 P(|0_x\rangle_A |1_x\rangle_B) + P(|1_x\rangle_A) \otimes \mathbf{1}_{\text{ex}}.$$

Similarly, n_{fil} corresponds to

$$M_{\text{fil}} \equiv \mathbf{1}_A \otimes \sum_j A_j^\dagger A_j = \mathbf{1}_A \otimes (c_\beta^2 P(|1_x\rangle_B) + s_\beta^2 P(|0_x\rangle_B) + \mathbf{1}_{\text{ex}}).$$

>From these forms, we notice that n_{ph} and n_{fil} are also obtained by the projection measurement

$$\{P_{00}, P_{11}, P_{10}, P_{01}, P(|0_x\rangle_A) \otimes \mathbf{1}_{\text{ex}}, P(|1_x\rangle_A) \otimes \mathbf{1}_{\text{ex}}\},$$

where $P_{ij} \equiv P(|i_x\rangle_A |j_x\rangle_B)$, followed by a classical procedure composed of Bernoulli trials. If we denote the results of the N projection measurements as $\{n_+(1 - \delta_+), n_+\delta_+, n_-(1 - \delta_-), n_-\delta_-, m_0, m_1\}$ in the same order, these numbers should be related to n_{ph} and n_{fil} as

$$|n_{\text{ph}} - m_1 - n_-[s_\beta^2(1 - \delta_-) + c_\beta^2 \delta_-]| \leq N\varepsilon \tag{4}$$

$$|n_{\text{fil}} - m_0 - m_1 - c_\beta^2(n_+\delta_+ + n_-\delta_-)$$
$$- s_\beta^2[n_+(1 - \delta_+) + n_-(1 - \delta_-)]| \leq N\varepsilon \tag{5}$$

with probability at least $1 - \exp(-2N\varepsilon^2)$. Since the marginal state ρ_A on K_A cannot be altered by Eve, the X-basis measurement on K_A is another Bernoulli trial. Since $\rho_A = c_\alpha^2 P(|0_x\rangle_A) + s_\alpha^2 P(|1_x\rangle_A)$, we have

$$|m_1 + n_+\delta_+ + n_-(1 - \delta_-) - s_\alpha^2 N| \leq N\varepsilon. \tag{6}$$

For the check pairs, n_{err} corresponds to

$$M_{\text{err}} \equiv P(|0_z\rangle_A) \otimes F_1 + P(|1_z\rangle_A) \otimes F_0 = (1/2)[P(|\Gamma_{11}\rangle) + P(|\Gamma_{01}\rangle) + \mathbf{1}_A \otimes \mathbf{1}_{\text{ex}}],$$

where we have introduced a basis $\{|\Gamma_{ij}\rangle\}_{i,j=0,1}$ of $K_A \otimes K_B$ by

$$|\Gamma_{ij}\rangle \equiv c_\beta |i_x\rangle_A |j_x\rangle_B - (-1)^j s_\beta |(1 - i)_x\rangle_A |(1 - j)_x\rangle_B.$$

It implies that n_{err} could also be obtained by the global projection measurement $\{Q_{00}, Q_{11}, Q_{10}, Q_{01}, \mathbf{1}_A \otimes \mathbf{1}_{\text{ex}}\}$, where $Q_{ij} \equiv P(|\Gamma_{ij}\rangle)$, followed by Bernoulli trials. If we write the results of N projection measurements as $\{n'_+(1 - \delta'_+), n'_+\delta'_+, n'_-(1 - \delta'_-), n'_-\delta'_-, m\}$, we obtain

$$|n_{\text{err}} - (n'_+\delta'_+ + n'_-\delta'_- + m)/2| \leq N\varepsilon. \tag{7}$$

If we compare the projection measurements on the data pairs and the check pairs, we further notice that n_+ and n'_+ are the results of an identical measurement, namely, projection onto the space H_+ spanned by $\{|0_x\rangle_A|0_x\rangle_B, |1_x\rangle_A|1_x\rangle_B\}$. We can thus apply the classical probability estimate. δ_+ and δ'_+ comes from projection to nonorthogonal states. For such a case, it was shown[9] that combination (δ_+, δ'_+) is exponentially rare unless there exists a state ρ on H_+ satisfying $\text{Tr}[\rho P(|1_x\rangle_A|1_x\rangle_B)] = \delta_+$ and $\text{Tr}[\rho P(|\Gamma_{11}\rangle)] = \delta'_+$. Using these arguments, we obtain

$$|n_\pm - n'_\pm| \leq N\varepsilon, \tag{8}$$

$$\delta'_\pm \geq c_\beta^2\delta_\pm + s_\beta^2(1 - \delta_\pm) - 2c_\beta s_\beta\sqrt{\delta_\pm(1 - \delta_\pm)} - \varepsilon. \tag{9}$$

Combining the above inequalities (4)–(9) with the following obvious relations

$$n_+ + n_- + m_0 + m_1 = N \tag{10}$$

$$n'_+ + n'_- + m = N, \tag{11}$$

we can calculate n_{phmax} as a function of given $(n_{\text{fil}}, n_{\text{err}})$.

4. Key gain

We are interested in the secret key gain in the limit $N \to \infty$. Let us define the quantities normalized by N such as $\hat{n}_{\text{ph}} \equiv n_{\text{ph}}/N$ and $\hat{n}_{\text{err}} \equiv n_{\text{err}}/N$, for example. The rate of the final key is given[5,7,16] by

$$G = \hat{n}_{\text{fil}}\left[1 - h\left(\frac{\hat{n}_{\text{bitmax}}}{\hat{n}_{\text{fil}}}\right) - h\left(\frac{\hat{n}_{\text{phmax}}}{\hat{n}_{\text{fil}}}\right)\right], \tag{12}$$

where $h(x) \equiv -x\log_2 x - (1 - x)\log_2(1 - x)$. In the limit of $N \to \infty$, we can set $\varepsilon = 0$ in the derived inequalities in the last section. Then, we can determine $\hat{n}_{\text{bitmax}}$ and $\hat{n}_{\text{phmax}}$ for given values of $\hat{n}_{\text{fil}}, \hat{n}_{\text{err}}$ in the following way.

First, from Eq. (3), we have

$$\hat{n}_{\text{bitmax}} = \hat{n}_{\text{err}}. \tag{13}$$

>From other relations, we can derive the following set of relations:

$$2\hat{n}_{\text{err}} \geq \hat{n}_{\text{fil}} - 2c_\beta s_\beta\left[\hat{n}_+\sqrt{\delta_+(1 - \delta_+)}\right.$$
$$\left. + \hat{n}_-\sqrt{\delta_-(1 - \delta_-)}\right] \tag{14}$$

$$\hat{n}_{ph} = \hat{m}_1 + \hat{n}_-[s_\beta^2(1-\delta_-)+c_\beta^2\delta_-] \tag{15}$$

$$\hat{n}_{fil} = \hat{m}_0 + \hat{m}_1 + c_\beta^2(\hat{n}_+\delta_+ + \hat{n}_-\delta_-)$$
$$+ s_\beta^2[\hat{n}_+(1-\delta_+)+\hat{n}_-(1-\delta_-)] \tag{16}$$

$$\hat{m}_1 + \hat{n}_+\delta_+ + \hat{n}_-(1-\delta_-) = s_\alpha^2 \tag{17}$$

$$\hat{n}_+ + \hat{n}_- + \hat{m}_0 + \hat{m}_1 = 1. \tag{18}$$

Then we can eliminate $\hat{n}_\pm, \delta_\pm$ in the first inequality by using the subsequent four equalities, obtaining an inequality of form $\hat{n}_{err} \geq f(\hat{n}_{fil}, \hat{n}_{ph}, \hat{m}_0, \hat{m}_1)$. Hence we can calculate the minimum $\hat{n}_{errmin}$ of $\hat{n}_{err}$ as a function of $\hat{n}_{fil}, \hat{n}_{ph}$ as

$$\hat{n}_{errmin}(\hat{n}_{fil}, \hat{n}_{ph}) = \min_{\hat{m}_0, \hat{m}_1} f(\hat{n}_{fil}, \hat{n}_{ph}, \hat{m}_0, \hat{m}_1).$$

Then we can determine $\hat{n}_{phmax}$ by solving $\hat{n}_{err} = \hat{n}_{errmin}(\hat{n}_{fil}, \hat{n}_{phmax})$.

5. High-loss limit

In this section, we consider the key rate in the limit of a high loss in the transmission, $\eta \to 0$ (examples of the key rate for low-loss cases were given elsewhere[17]). Since the observed parameters $\hat{n}_{fil}$ and $\hat{n}_{err}$ both go to zero in this limit, let us define normalized quantities instead, which stay nonzero. One is the error rate in the conclusive bits, defined by $r_{err} \equiv \hat{n}_{err}/\hat{n}_{fil}$. We similarly define the phase error rate $r_{ph} \equiv \hat{n}_{ph}/\hat{n}_{fil}$. Since $\hat{n}_{fil}$ goes to zero in proportion to η, we could normalized it by η. But here we introduce the quantity

$$\hat{n}_{fil0} := (1 - e^{-4\eta|\alpha|^2})/2, \tag{19}$$

which is the rate of obtaining conclusive outcomes that would be observed if the channel was perfect except for the transmission η. Let us normalize $\hat{n}_{fil}$ by this quantity, as $s_{fil} \equiv \hat{n}_{fil}/\hat{n}_{fil0}$. We further introduce $s_0 \equiv m_0/\hat{n}_{fil0}$, $s_1 \equiv m_1/\hat{n}_{fil0}$, and $s_\pm \equiv \delta_\pm/\hat{n}_{fil0}$. Then, Eqs. (15)–(18) in the limit $\eta \to 0$ become

$$\hat{n}_+ = c_\alpha^2 \tag{20}$$

$$\hat{n}_- = s_\alpha^2 \tag{21}$$

$$s_+ = \frac{s_{fil}(1-r_{ph})-s_0}{c_\alpha^2} - \frac{1}{2} \tag{22}$$

$$s_- = \frac{s_{fil}r_{ph}-s_1}{s_\alpha^2} - \frac{1}{2}. \tag{23}$$

The inequality (14) is also rewritten as

$$r_{err} \geq \frac{1}{2} - \frac{c_\alpha^2\sqrt{s_+}+s_\alpha^2\sqrt{s_-}}{\sqrt{2}s_{fil}} \tag{24}$$

As stated in the last section, $r_{phmax} \equiv \hat{n}_{phmax}/\hat{n}_{fil}$ can be obtained by taking the minimum of the right hand side of Eq. (24) over s_0 and s_1, which occurs when $s_0 = s_1 = 0$. Hence we arrive at the equation that implicitly determines r_{phmax} as a function of s_{fil} and r_{err}:

$$r_{err} = \frac{1}{2} - \frac{1}{2s_{fil}} \left(c_\alpha \sqrt{2s_{fil}(1 - r_{phmax}) - c_\alpha^2} \right.$$
$$\left. + s_\alpha \sqrt{2s_{fil}r_{phmax} - s_\alpha^2} \right). \tag{25}$$

The key rate in the same limit is given by

$$G/\eta \rightarrow 2|\alpha|^2 s_{fil} \left[1 - h(r_{err}) - h(r_{phmax}) \right] \tag{26}$$

which shows that, in the high loss limit, the key rate is proportional to the transmission η as long as the observed parameters r_{err} and s_{fil} give a positive value for the right hand side.

For example, in an ideal case where $r_{err} = 0, s_{fil} = 1$, we can solve Eq. (25) to obtain $r_{phmax} = s_\alpha^2$. Then, the key rate is

$$G/\eta \rightarrow 2|\alpha|^2 \left[1 - h \left(\frac{1 - e^{-2|\alpha|^2}}{2} \right) \right], \tag{27}$$

where the right hand side shows the trade off between the raw key rate and the information leak to Eve. It is optimized for $|\alpha|^2 = 0.23$, achieving $G \sim 0.14\eta$. Considering the use of conventional lasers, this result is not so bad when compared to the rate $G = 0.5\eta$ for the BB84 protocol with an ideal single-photon source.

The threshold error rate, above which the key rate vanishes, can be determined by the relation

$$1 - h(r_{err}) - h(r_{phmax}) = 0. \tag{28}$$

In general, the threshold error rate gets larger when the amplitude $|\alpha|$ is smaller. In the limit of $|\alpha|^2 \rightarrow 0$, we can explicitly write r_{phmax} as

$$r_{phmax} = 2s_{fil}r_{err}(1 - r_{err}) - \frac{(1 - s_{fil})^2}{2s_{fil}}. \tag{29}$$

Substituting this into Eq.(28), we can determine the threshold error rate as a function of s_{fil}, which is shown in Fig. 2. It is seen that as long as the error rate is smaller than $\sim 7.6\%$, we have a positive key gain. We also notice that in some region the key gain is positive for larger error rates, but we should note that such cases occur only when we take an eccentric model of noises in the transmission.

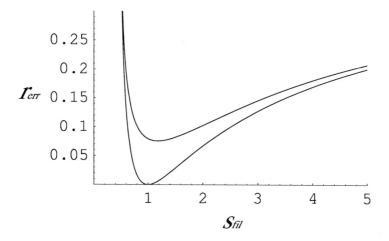

Figure 2. The threshold error rate (the upper curve). The region below the lower curve is not allowed physically.

6. Summary

In summary, we have shown that by encoding on the phase of a weak coherent pulse relative to a strong reference pulse, we can achieve a key rate of $O(\eta)$ with unconditional security, which is an advantage over the coherent-state BB84. The key rate is positive up to the error rate $\sim 7.6\%$. There are several proposals[18,19] to improve the performance of the coherent-state BB84, and their unconditional security[20] is an interesting problem. The security of the original B92, which uses only one LO, is also interesting since the relation between the amplitude of the reference pulse and the security will show up more tightly.

Acknowledgments

The author thanks N. Imoto, H.-K. Lo, D. Mayers, J. Preskill, K. Tamaki, and especially N. Lütkenhaus for helpful discussions. This work was supported by 21st Century COE Program by the Japan Society for the Promotion of Science.

References

1. D. Mayers, *Lect. Notes Comput. Sci.* **1109**, 343 (1996).
2. C. H. Bennett and G. Brassard, in *Proceeding of the IEEE International Conference on Computers, Systems, and Signal Processing, Bangalore, India*, pp. 175–179, IEEE, New York, 1984.
3. E. Biham, M. Boyer, P. O. Boykin, T. Mor, and V. Roychowdhury, *quant-ph/9912053*.
4. H. K. Lo and H. F. Chau, *Science* **283**, 2050 (1999).

5. P. W. Shor and J. Preskill, *Phys. Rev. Lett.* **85**, 441 (2000).

6. H. Inamori, N. Lütkenhaus, and D. Mayers, *quant-ph/0107017* .

7. D. Gottesman, H. K. Lo, N. Lütkenhaus, and J. Preskill, *quant-ph/0212066*.

8. M. Koashi and J. Preskill, *Phys. Rev. Lett.* **90**, 057902 (2003).

9. K. Tamaki, M. Koashi, and N. Imoto, *Phys. Rev. Lett.* **90**, 167904 (2003).

10. K. Tamaki and N. Lütkenhaus, *Phys. Rev. A* **69**, 032316 (2004).

11. M. Christandl, A. Ekert, and R. Renner, *quant-ph/0402131*.

12. N. Lütkenhaus, *Phys. Rev. A* **61**, 052304 (2000).

13. G. Brassard, N. Lütkenhaus, T. Mor, and B. C. Sanders, *Phys. Rev. Lett.* **85**, 1330 (2000).

14. C. H. Bennett, *Phys. Rev. Lett* **68**, 3121 (1992).

15. C. H. Bennett, D. P. DiVincenzo, J. A. Smolin, and W. K. Wootters, *Phys. Rev. A* **54**, 3824 (1996).

16. M. Hamada, *quant-ph/0308039* .

17. M. Koashi, *Phys. Rev. Lett.* **93**, 120501 (2004).

18. W. Y. Hwang, *Phys. Rev. Lett.* **91**, 057901 (2003).

19. V. Scarani, A. Acín, and N. Gisin, *Phys. Rev. Lett.* **92**, 057901 (2004).

20. H. K. Lo, X. Ma, and K. Chen, *quant-ph/0411004*.

LIMITATION ON THE ACCESSIBLE INFORMATION FOR QUANTUM CHANNELS WITH INEFFICIENT MEASUREMENTS

KURT JACOBS

Centre for Quantum Computer Technology, Centre for Quantum Dynamics,
School of Science, Griffith University, Nathan 4111, Australia

To transmit classical information using a quantum system, the sender prepares the system in one of a set of possible states and sends it to the receiver. The receiver then makes a measurement on the system to obtain information about the senders choice of state. The amount of information which is accessible to the receiver depends upon the encoding and the measurement. Here we derive a bound on this information which generalizes the bound derived by Schumacher, Westmoreland and Wootters [Schumacher, Westmoreland and Wootters, Phys. Rev. Lett. **76**, 3452 (1996)] to include inefficient measurements, and thus all quantum operations. This also allows us to obtain a generalization of a bound derived by Hall [Hall, Phys. Rev. A **55**, 100 (1997)], and to show that the average reduction in the von Neumann entropy which accompanies a measurement is concave in the initial state, for all quantum operations.

1. Introduction

Here we are concerned with the scenario in which one observer (A, or Alice) sends classical information to B (or Bob), by encoding this information in a quantum system. To do this, Alice and Bob agree on a set of states, $\{\rho_i\}$ to use as an "alphabet" for communication. Alice prepares the system in one of these states, and sends it to Bob, who measures the system so as to determine as best he can which state alice has sent. Bob's measurement, being a quantum measurement,[1,2] is described by a set of operators A_j where each j corresponds to one of the possible measurement results. Since Bob does not know which state Alice has prepared prior to making his measurement, his initial state for the system is given by $\sum_i P(i)\rho_i$ where $P(i)$ is the probability that alice will send the message i. The probability that Bob obtains outcome j is given by $P(j) = \text{Tr}[A_j^\dagger A_j \rho]$.[a] The conditional probability that Bob obtains result j *given* that Alice has prepared state i is $P(j|i) = \text{Tr}[A_j^\dagger A_j \rho_i]$. Upon obtaining the result j Bob's state of knowledge for

---

[a]While it is an abuse of notation to denote the ensemble probabilities by $P(i)$, and the (in general unrelated) outcome probabilities by $P(j)$, we use it systematically throughout, since we feel it keeps the notation simpler, and thus ultimately clearer.

the system becomes $\sigma_j = A_j \rho A_j^\dagger / P(j)$. We note that the set of states along with their respective probabilities is referred to as an *ensemble*, and we will refer to ρ as the *ensemble state*.

The amount of information which Bob obtains from Alice (being information about which value of i Alice has chosen) is given by the *mutual information* between the random variable I who's values are i (note that i is random from Bob's point of view), and the random variable j. The mutual information is

$$M(I : J) = H[P(i)] - \sum_j P(j) H[P(i|j)],\tag{1}$$

where $H[P(i)] = -\sum_i P(i) \ln P(i)$ is the Shannon entropy of the probability distribution $P(i)$. The expression for the mutual information is easy to understand. It is the difference between the entropy of Bob's state-of-knowledge regarding i (given by $P(i)$) before he makes the measurement and his state-of-knowledge after he makes the measurement (given by $P(i|j)$) averaged over the possible measurement results. Note that in general Bob's information will not be complete, in that there will still be some entropy left in his final probability distribution $P(i|j)$. However the reason that the mutual information is so useful is because of Shannon's noisy coding theorem. This states that, in the limit that the transmission procedure (referred to as "the channel") is used many times, it is possible to code information across multiple uses of the channel such that Bob does get all the information that was encoded. Further, in the case of classical channels the amount of information that can be sent reliably per use of the channel is precisely the mutual information $M(I,J)$.[3] In the quantum case $M(I : J)$ is just a lower bound on the reliable transmission rate, because it is possible to make joint measurements across multiple uses of the channel (that is, jointly measure more than one system at a time) and obtain a higher rate.[4,5]

The celebrated Holevo bound, conjectured by Gordon[6] and Levitin,[7,8] and proved by Holevo in 1973,[9] gives a bound on the mutual information for a given encoding ensemble. This states that, if the receiver has the ability to perform any measurement, then

$$M(I:J) \le \chi(\varepsilon) \equiv S(\rho) - \sum_i P(i) S(\rho_i)\tag{2}$$

The quantity on the RHS is called the Holevo χ quantity for the ensemble $\varepsilon = \{\rho_i, P(i)\}$.

In 1996 Schumacher, Westmoreland and Wootters (SWW) derived a bound for the case in which the measurement is also fixed and may in general be incomplete.[10] This bound is more stringent that the Holevo bound when the measurement is incomplete, and reduces to the Holevo bound for complete measurements. Incomplete measurements are measurements in which the measurement operators A_j are not all rank-1, so that in general the measurement does not leave the system in a pure state. As a result, if the system starts in some ensemble of states, then it will, in general, remain in a (different) ensemble of states after the measurement. The SWW bound is[10]

$$M(I:J) \leq \chi - \sum_j P(j)\chi_j, \tag{3}$$

where $P(j)$ is the probability of outcome j, and χ_j is the Holevo quantity for the ensemble that the system remains in (from the point of view of the receiver), given outcome j. This bound can be at least partially understood by noting that if the system still remains in some ensemble of possible states after the measurement, then future measurements can potentially extract further information about the encoding, and so the information obtained by the first measurement must therefore be less than the maximum possible by at least by this amount. What the SWW bound tells us is that the bound on the information is reduced not only by the amount of information which could be further extracted after outcome j has been obtained, but by the Holevo bound on this information, χ_j.

The most general kind of quantum measurement can be both inefficient as well as incomplete. An inefficient measurement is one in which the observer only has partial knowledge of the outcome of the measurement. Any incomplete measurement can be described by labeling the measurement operators by two indices, and giving the observer access only to one of them when the measurement has been made.[b] In the this case Bob must average the final states over the values of the variable he does not know. Such measurements are called inefficient because they first arose in the context of inefficient photo-detection.[11]

We show here that the SWW theorem is also true for inefficient measurements, and thus for all quantum measurements. We also consider a bound derived by Hall, which states that for efficient measurements the mutual information is bounded by the average reduction in the von Nuemann entropy of the system caused by the measurement.[12,13] This is no-longer true for inefficient measurements, but the extension of the SWW theorem to inefficient measurements allows us to obtain as a corollary a generalization of Hall's bound which is valid for all measurements.

---

[b]If the observer has only partial information about the outcome of a measurement, then if we label the outcomes by n (with associated measurement operators B_n), the most general situation is one in which the observer knows instead the value of a second variable m, where m is related to n by an arbitrary conditional probability $P(m|n)$. This general case is encompassed by the two-index formulation we use in the text. To see this, one sets $k = n$, $j = m$ and chooses $A_{nm}(\equiv A_{kj}) = \alpha_{nm}B_n$. Then by giving the observer complete knowledge of j, and no knowledge of k, we reproduce precisely the general case described above by choosing α_{nm} so that $|\alpha_{nm}|^2 = P(m|n)$.

We also obtain a second corollary, which is that the average reduction in the von Neumann entropy is concave in the initial state. This has recently proved to be useful in the quantification of the asymmetry of a state with regard to a reference frame.[14]

We note that the average reduction in the von Neumann entropy is a useful quantity in quantum feedback control. In feedback control, one continually observers a system (a process which can be though of as a sequence of many weak measurements) and uses the information gained, in real-time, to apply forces by adjusting the system Hamiltonian, and thus applying a sequence of unitary operations to the system.[15,16,17,18,19] One's ability to control a system in this manner, assuming that there is no restriction upon the unitaries that may be applied, is determined solely by the entropy of the system.[20] The primary role of measurement in feedback control is therefore to reduce this entropy. As such the average reduction in von Neumann entropy provides a ranking of the effectiveness of different measurements for feedback control, other things being equal. Further details regarding quantum feedback control and von Neumann entropy can be found in reference 20.

In the next section we extend the SWW bound to inefficient measurements, and in Section 3 we present the corollaries regarding Hall's bound and the von Neumann entropy. Section 4 concludes with a brief summary of the relationships between the various information bounds.

2. Extending the SWW Bound to Inefficient Measurements

To prove their theorem, SWW considered a situation consisting of three subsystems, the system to be measured, a second system through which the measurement is accomplished, and a third system playing the role of an environment. It turns out that to extend the SWW theorem to inefficient measurements we can employ the same method as used by SWW, but with the addition of a forth system which allows us to include the inefficiency of the measurement.

Theorem 1. *For a communication protocol in which the sender encodes information in a quantum system using the ensemble $\varepsilon = \{P(i), \rho_i\}$, and the receiver makes a measurement described by the operators A_{kj} ($\sum_{kj} A_{kj}^\dagger A_{kj} = 1$), where the measurement is in general inefficient so that the receiver knows j but not k, then the mutual information, $M(I:J)$, is bounded such that*

$$M(I:J) \le \chi - \sum_j P(j)\chi_j, \qquad (4)$$

where $P(j)$ is the overall probability for outcome j, $\chi = S(\rho) - \sum_i P(i)S(\rho_i)$ is the

Holevo quantity for the initial ensemble and

$$\chi_j = S(\sigma_j) - \sum P(i|j)S(\sigma_{j|i}), \tag{5}$$

is the Holevo quantity for the ensemble, ε_j, that remains from the point of view of the receiver once the measurement has been made (so that the receiver has learned the value of j, but not the value of k). The receiver's overall final state is

$$\sigma_j = \frac{\sum_k A_{kj}\rho A_{kj}^\dagger}{P(j)} = \sum_{ik} P(i,k|j)\sigma_{kj|i}, \tag{6}$$

where $P(i,k|j)$ is the probability for both i and outcome k given j, and $\sigma_{kj|i}$ is the final state that results given the initial state ρ_i, and both outcomes j and k. The remaining ensemble is $\varepsilon_j = \{P(j|i), \sigma_{j|i}\}$, where

$$\sigma_{j|i} = \sum_k P(k|j,i)\sigma_{kj|i} = \frac{\sum_k A_{kj}\rho_i A_{kj}^\dagger}{P(j|i)}, \tag{7}$$

and where $P(k|j,i)$ is the probability for outcome k given j and the initial state ρ_i.

Proof. To begin with it is useful to collect a number of key facts. The first is that any efficient measurement on a system Q, described by $N = N_1 N_2$ operators, A_{kj}, ($j = 1, \ldots, N_1$ and $k = 1, \ldots, N_2$) can be obtained by performing a unitary operation between Q and an auxiliary system A of dimension N, and then making a von Neumann measurement on A.[21,2] If the initial state of Q is $\rho^{(Q)}$, then the final joint state of A and Q after the von Neumann measurement is

$$\sigma^{(AQ)} = |kj\rangle\langle kj|^{(A)} \otimes \frac{A_{kj}\rho^{(Q)}A_{kj}^\dagger}{P(k,j)}, \tag{8}$$

where $|kj\rangle$ is the state of A selected by the von Neumann measurement. The second fact is that the state which results from discarding all information about the measurement outcomes k and j can be obtained by performing a unitary operation between A and another system E which perfectly correlates the states $|kj\rangle$ of A with orthogonal states of E, and then tracing out E. The final key fact we require is a result proven by SWW,[10] which is that the Holevo χ quantity is non-increasing under partial trace. That is, if we have two quantum systems A and B, and an ensemble of states $\rho_i^{(AB)}$ with associated probabilities P_i, then

$$\chi^{(A)} = S(\rho^{(A)}) - \sum_i S(\rho_i^{(A)}) \le S(\rho^{(AB)}) - \sum_i S(\rho_i^{(AB)}) = \chi^{(AB)}, \tag{9}$$

where $\rho_i^{(A)} = \text{Tr}_B[\rho_i^{(AB)}]$. To prove this result SWW use strong subadditivity.[c]

---

[c]The original proof of strong subadditivity is given in references 22, 23 and 24. In addition, simpler proofs of strong subadditivity are given in Petz[25] (which is described in Ref. 26) and Ref. 27 .

We now encode information in system Q using the ensemble ε, and consider the joint system which consists of the three systems described above, Q, A, and E, and a forth system M that has dimension N_1. We now start with A, E and M in pure states, so that the Holevo quantity for the joint system is $\chi^{(QAEM)} = \chi^{(Q)}$. We then perform the required unitary operation between Q and A, and a unitary operation between A and E which perfectly correlates the states $|kj\rangle^{(A)}$ of A with orthogonal states of E. Unitary operations do not change the Holevo quantity. Then we trace over E, so that we are left with the state

$$|\psi\rangle\langle\psi|^{(M)} \otimes \sum_{jk} P(k,j)|k,j\rangle\langle k,j|^{(A)} \otimes \frac{A_{kj}\rho^{(Q)}A_{kj}^\dagger}{P(k,j)}. \tag{10}$$

After the two unitaries and the partial trace over E, the Holevo quantity for the remaining systems, which we will denote by $\chi'^{(QAM)}$, satisfies $\chi'^{(QAM)} \leq \chi^{(QAEM)} = \chi^{(Q)}$. We now perform one more unitary operation, this time between M and A, so that we correlate the states of M, which we denote by $|j\rangle\langle j|^{(M)}$ with the second index of the states of A, giving

$$\sum_j |j\rangle\langle j|^{(M)} \otimes \sum_k P(k,j)|k,j\rangle\langle k,j|^{(A)} \otimes \sigma_{kj}^{(Q)} \tag{11}$$

where $\sigma_{kj}^{(Q)} = A_{kj}\rho^{(Q)}A_{kj}^\dagger/P(k,j)$ is the final state resulting from knowing both outcomes k and j, with no knowledge of the initial choice of i. Finally we trace out A, leaving us with the state

$$\sigma^{(QM)} = \sum_j |j\rangle\langle j|^{(M)} \otimes \sum_k P(k,j)\sigma_{kj}^{(Q)} \tag{12}$$

After this final unitary, and the partial trace over A, the Holevo quantity for the remaining systems Q and M, which we will denote by $\chi''^{(QM)}$, satisfies $\chi''^{(QM)} \leq \chi'^{(QAM)} \leq \chi^{(Q)}$. We have gone through the above process using the initial state ρ, but we could just as easily have started with any of the initial states, ρ_i, in the ensemble, and we will denote the final states which we obtain using the initial state ρ_i as $\sigma_i^{(QM)}$. Calculating $\chi''^{(QM)}$ we have

$$\chi''^{(QM)} = S(\sigma^{(QM)}) - \sum_i P(i)S(\sigma_i^{(QM)})$$

$$= H[J] - \sum_i P(i)H[J|i] + \sum_j P(j)\left[S(\sigma_j) - \sum_i P(i|j)\sigma_{j|i}\right] \tag{13}$$

$$= M(J:I) + \sum_j P(j)\chi_j^{(Q)} \leq \chi^{(Q)}. \tag{14}$$

Rearranging this expression gives the desired result. □

3. Two corollaries regarding entropy reduction

Hall's bound states that[12,13]

$$M(I:J) \leq \langle \Delta S(\rho) \rangle \qquad (15)$$

is always true for efficient measurements. However, this cannot always be true for inefficient measurements because in this case the entropy reduction $\langle \Delta S(\rho) \rangle$ can be *negative*, whereas the mutual information is always positive. We can obtain the generalization of Hall's bound for inefficient measurements by rewriting the information bound derived above using the fact that $P(i|j)P(j) = P(j|i)P(i)$. This gives

$$M(I:J) \leq \langle \Delta S(\rho) \rangle - \sum_i P(i) \langle \Delta S(\rho_i) \rangle \qquad (16)$$

where $\rho = \sum_i P_i \rho_i$. This also tells us that if the entropy of the system does increase on average when the measurement is made, the average increase in the entropy for each of the coding states ρ_i is always *more* that this by at least the mutual information. For efficient measurements this expression generates Hall's bound because $\langle \Delta S(\rho) \rangle$ is always positive for such measurements.[28,29,30]

The second result that we obtain from Eq.(16) is that, because the mutual information is nonnegative, we have

$$\langle \Delta S(\rho) \rangle \geq \sum_i P(i) \langle \Delta S(\rho_i) \rangle. \qquad (17)$$

That is, the reduction in the von Neumann entropy is concave in the initial state. This parallels the fact that the mutual information is also concave in the initial state. The fact that this is true for inefficient measurements, and not just efficient measurements, means that, once we have made an efficient measurement, it remains true no matter what information we choose to throw away regarding the measurement results. This is true even though, the LHS of the above inequality can certainly decrease by *more* than the RHS when information is thrown away.

4. conclusion

In this work we have shown that the bound on the mutual information derived by Schumacher, Westmoreland and Wootters, also holds for inefficient measurements, and thus all quantum operations. The SWW implies both the Holevo bound and Hall's bound, which are complementary to each other: The Holevo bound applies when the sender's encoding is fixed and the receiver can make any measurement, and Hall's bound gives a bound in the opposite situation. In fact, they are even more closely related: the relationship between them is the result of the duality between states and measurements which was pointed out by Hall.[12] They therefore have very similar forms — while Holevo's bound involves the ensemble state and the initial encoding states, Hall's bound involves the ensemble state and the *final* states. In extending the SWW bound to inefficient measurements we obtain a generalization of Hall's bound which is valid for all quantum operations. In addition, this allows us to show that the reduction in the von Neumann entropy of a system during a measurement is concave in the initial state.

Acknowledgments

The author would like to thank Gerard Jungman, Howard Barnum, Howard Wiseman, Terry Rudolph and Michael Hall for helpful discussions.

Note added: Shortly before publication the author became aware of the similar work by Barchielli and Lupieri.[31,32]

References

1. M. A. Nielsen and I. L. Chuang, *Quantum Computation and Quantum Information* (Cambridge University Press, 2000).
2. B. Schumacher, Phys. Rev. A **54**, 2614 (1996).
3. C. E. Shannon and W. Weaver, *The Mathematical Theory of Communication* (University of Illinois Press, Chicago, 1963).
4. A. S. Holevo, IEEE Trans. Inf. Theory **44**, 269 (1998).
5. B. Schumacher and M. D. Westmoreland, Phys. Rev. A **56**, 131 (1997).
6. J. P. Gordon, in *Quantum Electronics and Coherent Light, Proceedings of the International School of Physics 'Enrico Fermi' XXXI*, edited by P. Miles (Accademic Press, New York, 1964).
7. L. B. Levitin, in *Proceedings of the All-Union Conference on Information Complexity and Control in Quantum Physics, Sec. II*, edited by P. Miles (Mockva-Tashkent, Tashkent, 1969).
8. L. B. Levitin, in *Information, Complexity and Control in Quantum Physics*, edited by A. Blaquiere, S. Diner, and G. Lochak (Springer-Verlag, New York, 1987).
9. A. S. Holevo, Probl. Peredachi Inf. [Probl. Inf. Transm. (USSR) **9**, 177 (1973)] **9**, 3 (1973).
10. B. Schumacher, M. Westmoreland, and W. K. Wootters, Phys. Rev. Lett. **76**, 3452 (1996).
11. H. M. Wiseman and G. J. Milburn, Phys. Rev. A **47**, 642 (1993).
12. M. J. W. Hall, Phys. Rev. A **55**, 100 (1997).
13. K. Jacobs, Phys. Rev. A **68**, 054302 (2003).
14. J. A. Vaccaro, F. Anselmi, H. M. Wiseman, and K. Jacobs, *Complementarity between extractable mechanical work, accessible entanglement, and ability to act as a reference frame, under arbitrary superselection rules* (Eprint: quant-ph/0501121).
15. V. P. Belavkin, in *Information, Complexity and Control in Quantum Physics*, edited by A. Blaquiere, S. Diner, and G. Lochak (Springer-Verlag, New York, 1987).

16. V. P. Belavkin, Rep. Math. Phys. **43**, 405 (1999).
17. A. C. Doherty and K. Jacobs, Phys. Rev. A **60**, 2700 (1999).
18. A. C. Doherty, S. Habib, K. Jacobs, H. Mabuchi, and S. M. Tan, Phys. Rev. A **62**, 012105 (2000).
19. M. R. James, Phys. Rev. A **69**, 032108 (2004).
20. A. C. Doherty, K. Jacobs, and G. Jungman, Phys. Rev. A **63**, 062306 (2001).
21. K. Kraus, *States, Effects and Operations: Fundamental Notions of Quantum Theory, Lecture Notes in Physic Vol. 190* (Springer-Verlag, Berlin, 1983).
22. E. H. Lieb, Ad. Math. **11**, 267 (1973).
23. E. H. Lieb and M. B. Ruskai, Phys. Rev. Lett. **30**, 434 (1973a).
24. E. H. Lieb and M. B. Ruskai, J. Math. Phys. **14**, 1938 (1973b).
25. D. Petz, Rep. on Math. Phys. **23**, 57 (1986).
26. M. A. Nielsen and D. Petz, *A simple proof of the strong subadditivity inequality* (Eprint: quant-ph/0408130).
27. M. B. Ruskai, J. Math. Phys. **43**, 4358 (2002).
28. M. Ozawa, J. Math. Phys. **27**, 759 (1986).
29. M. A. Nielsen, Phys. Rev. A **63**, 022114 (2001).
30. C. A. Fuchs and K. Jacobs, Phys. Rev. A **63**, 062305 (2001).
31. A. Barchielli and G. Lupieri, *Instruments and channels in quantum information theory* (Eprint: quant-ph/0409019).
32. A. Barchielli and G. Lupieri, *Instruments and mutual entropies in quantum information* (Eprint: quant-ph/0412116).

CODES FOR KEY GENERATION
IN QUANTUM CRYPTOGRAPHY

BERTHOLD-GEORG ENGLERT

Department of Physics, National University of Singapore, Singapore 117542
phyebg@nus.edu.sg

FANG-WEI FU[*]

Temasek Laboratories, National University of Singapore, Singapore 117508
tslfufw@nus.edu.sg

HARALD NIEDERREITER AND CHAOPING XING

Department of Mathematics, National University of Singapore, Singapore 117543
nied@math.nus.edu.sg, matxcp@nus.edu.sg

As an alternative to the usual key generation by two-way communication in schemes for quantum cryptography, we consider codes for key generation by one-way communication. We study codes that could be applied to the raw key sequences that are ideally obtained in recently proposed scenarios for quantum key distribution, which can be regarded as communication through symmetric four-letter channels.

Keywords: Error correcting codes, linear codes, quantum key distribution

1. Introduction

In a recently proposed protocol for quantum key distribution,[1,2] Alice sends uncorrelated qubits through a quantum channel to Bob. Under ideal circumstances, the channel is noiseless, and then the situation is as follows.

Alice prepares each qubit in one of four states — labeled A, B, C, and D, respectively, and chosen at random — and Bob detects each qubit in one of four states that are labeled correspondingly. The set-up has the peculiar feature that Bob *never* obtains the letter that specifies the state prepared by Alice. Rather, he

---

[*]on leave from the department of mathematics, nankai university, tianjin 300071, p. r. china

always gets one of the other three letters, whereby the laws of quantum physics ensure that the outcome is truly random, and each possibility occurs equally likely.

These physical laws also prevent any third party, eavesdropper Eve, from acquiring information about Alice's or Bob's letters. Therefore, they can exploit the correlations between their letters to generate a private cryptographic key, which they can then use for the secure encryption of a message.

The key generation is a crucial step. Two different procedures are described in Refs. 1 and 2, with respective efficiencies of $\frac{1}{3}$ and $\frac{2}{5}$ key bits per letter. Both procedures rely on *two-way* communication between Alice and Bob. By contrast, it is our objective here to study codes for the key generation by *one-way* communication.

After the exchange of many qubits through the quantum channel, Alice and Bob have random sequences of the four letters, such that corresponding letters are never the same, while each of the twelve pairs of different letters occurs one-twelfth of the time, with no correlations between the pairs. Alice sends a code word to Bob by telling him, through a public channel, the positions at which the letters appear in her sequence — such as "3rd letter, then 14th, 15th, 92nd, and 65th" for a particular five-letter word. Bob forms the received word from his corresponding letters, and then decodes.

The public communication does not leak any useful information to Eve. Thus, if Alice chooses a random sequence of code words, each word being equally likely, as she will do, Eve knows nothing about Alice's words. She also knows nothing about Bob's decoded words, provided that Bob's decoding procedure does not favor some words at the expense of others. Accordingly, the sequence of words constitutes a privately shared key for secure classical communication between Alice and Bob.

There is a nonzero probability that Bob's received word is consistent with two or more words that Alice could have sent, so that the decoding will not be completely error-free. A good, practical code must, therefore, represent a compromise between (i) having not too many code words, (ii) an acceptable error rate, and (iii) a reasonable efficiency. Arguably the best compromise we report in Section 8 is code (3) of Example 1. It has 1024 words, an error rate of 0.6%, and an efficiency of $\frac{1}{4}$ key bits per letter.

As there is no fundamental reason why the key generation by one-way communication should be substantially less efficient than that by two-way communication, one expects that more efficient codes can be found. Therefore, the work reported here should be regarded as a first step, not as the final word on the matter.

It is worth mentioning that there is a very similar problem for the three-letter channel of Renes's "trine" scheme.[1] Further, the standard BB84 protocol[3] has a four-letter channel with quite different properties, for which codes for one-way key generation are not known. The same remark applies to the six-letter generalization[4] of BB84. In short, there is a whole class of coding problems that deserve attention.

2. Probability Distributions

The quantum protocols of Alice and Bob involve two random variables X and Y taking values in $\{A,B,C,D\}$. We have the following corresponding probability distributions with $x,y \in \{A,B,C,D\}$. The joint probability distribution of X and Y is given by

$$\Pr\{X = x, Y = y\} = \begin{cases} 0 & \text{if } y = x, \\ \dfrac{1}{12} & \text{if } y \neq x. \end{cases} \tag{1}$$

Accordingly, the marginal probability distributions of X and Y are

$$\Pr\{X = x\} = \frac{1}{4}, \qquad \Pr\{Y = y\} = \frac{1}{4}, \tag{2}$$

and the conditional probability distribution of Y with respect to X is

$$\Pr\{Y = y \mid X = x\} = \begin{cases} 0 & \text{if } y = x, \\ \dfrac{1}{3} & \text{if } y \neq x. \end{cases} \tag{3}$$

Now we compute the information-theoretic quantities entropy, conditional entropy, and mutual information[a] of the random variables X and Y. The entropy of Y is

$$H(Y) = -\sum_{y \in \{A,B,C,D\}} \Pr\{Y = y\} \log_2 \Pr\{Y = y\} = 2, \tag{4}$$

and for the conditional entropy of Y with respect to X we find

$$H(Y|X) = -\sum_{\substack{x,y \in \{A,B,C,D\} \\ y \neq x}} \Pr\{X = x, Y = y\} \log_2 \Pr\{Y = y|X = x\} = \log_2 3, \tag{5}$$

and we obtain

$$I(X;Y) = \sum_{\substack{x,y \in \{A,B,C,D\} \\ y \neq x}} \Pr\{X = x, Y = y\} \log_2 \frac{\Pr\{X = x, Y = y\}}{\Pr\{X = x\} \Pr\{Y = y\}}$$

$$= H(Y) - H(Y|X) = \log_2 \frac{4}{3} \tag{6}$$

for the mutual information of X and Y.

---

[a]For the definitions of these and other information-theoretic quantities see Ref. 5, for example.

3. Discrete Memoryless Channel

The information transmission from Alice to Bob can be described in information theory by a discrete memoryless channel.[a] This channel is characterized by the conditional probability distribution[a] of Y with respect to X,

$$Q(y|x) = \Pr\{Y = y \mid X = x\} = \begin{cases} 0 & \text{if } y = x, \\ \dfrac{1}{3} & \text{if } y \neq x, \end{cases} \tag{7}$$

where $x,\ y \in \{A,B,C,D\}$. The *channel capacity*[a] is defined by

$$c = \max_{P_X} I(X;Y) = \max_P I(P;Q), \tag{8}$$

where

$$I(P;Q) = \sum_{x,y} P(x)Q(y|x) \log_2 \frac{Q(y|x)}{\sum_{x'} P(x')Q(y|x')} \tag{9}$$

and the maximum is taken over all probability distributions P on $\{A,B,C,D\}$.

The channel defined by (7) is a symmetric channel (see Theorem 8.2.1 on p. 190 in Ref. 5). Hence, the capacity is attained by the uniform distribution on $\{A,B,C,D\}$, so that

$$c = \log_2 \frac{4}{3} \doteq 0.4150. \tag{10}$$

For any positive integer n, the nth extension of this discrete memoryless channel has the conditional probability distribution

$$Q^n(\mathbf{y}|\mathbf{x}) = \begin{cases} 0 & \text{if } y_i = x_i \text{ for some } i, \\ \dfrac{1}{3^n} & \text{if } y_i \neq x_i \text{ for all } i, \end{cases} \tag{11}$$

where $\mathbf{x} = (x_1, x_2, \ldots, x_n)$, $\mathbf{y} = (y_1, y_2, \ldots, y_n) \in \{A,B,C,D\}^n$ are n-letter words.

4. Codes for the Specific Channel

In this section, we discuss the design of codes and decoding methods for the specific channel introduced in Section 3.

Let $\mathbf{F}_4$ be the finite field with four elements. It is convenient to let A, B, C, D be represented respectively by the four elements 0, 1, a, b of $\mathbf{F}_4$ since we want to use linear codes for this specific channel. The addition and multiplication tables of $\mathbf{F}_4$ are as follows:

$$
\begin{array}{c|cccc}
+ & 0 & 1 & a & b \\
\hline
0 & 0 & 1 & a & b \\
1 & 1 & 0 & b & a \\
a & a & b & 0 & 1 \\
b & b & a & 1 & 0
\end{array}
\qquad
\begin{array}{c|cccc}
\times & 0 & 1 & a & b \\
\hline
0 & 0 & 0 & 0 & 0 \\
1 & 0 & 1 & a & b \\
a & 0 & a & b & 1 \\
b & 0 & b & 1 & a
\end{array}
\tag{12}
$$

Let $\mathbf{F}_4^n$ be the n-dimensional vector space over $\mathbf{F}_4$. For two vectors $\mathbf{x} = (x_1, x_2, \ldots, x_n) \in \mathbf{F}_4^n$, $\mathbf{y} = (y_1, y_2, \ldots, y_n) \in \mathbf{F}_4^n$, the *Hamming distance* $d(\mathbf{x}, \mathbf{y})$ between $\mathbf{x}$ and $\mathbf{y}$ is defined as the number of coordinates in which they differ,

$$
d(\mathbf{x}, \mathbf{y}) = \left| \{ i : x_i \neq y_i \} \right|. \tag{13}
$$

The *Hamming weight* $w(\mathbf{x})$ is the number of nonzero coordinates in $\mathbf{x}$,

$$
w(\mathbf{x}) = \left| \{ i : x_i \neq 0 \} \right|. \tag{14}
$$

A code of length n with $M \geq 2$ codewords is a subset C of $\mathbf{F}_4^n$,

$$
C = \{ \mathbf{c}_1, \mathbf{c}_2, \ldots, \mathbf{c}_M \}, \qquad \mathbf{c}_i \in \mathbf{F}_4^n. \tag{15}
$$

The *minimum distance* $d(C)$ of the code C is the minimum Hamming distance between two distinct codewords,

$$
d(C) = \min \{ d(\mathbf{x}, \mathbf{y}) : \mathbf{x}, \mathbf{y} \in C, \ \mathbf{x} \neq \mathbf{y} \}. \tag{16}
$$

We denote by

$$
L(\mathbf{c}_i) = \{ \mathbf{x} \in \mathbf{F}_4^n : d(\mathbf{x}, \mathbf{c}_i) = n \} \quad \text{for } i = 1, 2, \ldots, M \tag{17}
$$

the set of n-letter words that could be received if codeword $\mathbf{c}_i$ is sent. It is easy to see that for any $i, j, l \in \{ 1, 2, \ldots, M \}$ we have

$$
\left| L(\mathbf{c}_i) \right| = 3^n, \tag{18}
$$

$$
\left| L(\mathbf{c}_i) \cap L(\mathbf{c}_j) \right| \geq 2^n, \tag{19}
$$

$$
\left| L(\mathbf{c}_i) \cap L(\mathbf{c}_j) \cap L(\mathbf{c}_l) \right| \geq 1. \tag{20}
$$

Note, in particular, the significance of (20): For any three different codewords that could have been sent by Alice, there is at least one word received by Bob that is consistent with all three.

Further, it follows from (11) that

$$
Q^n(\mathbf{y} | \mathbf{x}) = \begin{cases} 0 & \text{if } \mathbf{y} \notin L(\mathbf{x}), \\[2mm] \dfrac{1}{3^n} & \text{if } \mathbf{y} \in L(\mathbf{x}). \end{cases} \tag{21}
$$

This means that if Alice sends $x \in F_4^n$ through this channel to Bob, then Bob receives $y \in L(x)$ with probability $1/3^n$.

Now we describe a decoding method for the code C by decoding regions, which exploits the significance of $L(c_i)$. The M subsets $D_1, D_2, \ldots, D_M$ of F_4^n are called *decoding regions* for the code C if they satisfy the following conditions:

$$
\begin{aligned}
&\text{(i) } D_i \subseteq L(c_i), \quad i = 1, 2, \ldots, M; \\
&\text{(ii) } D_i \cap D_j = \emptyset, \quad i \neq j; \\
&\text{(iii) } \bigcup_{i=1}^{M} D_i = \bigcup_{i=1}^{M} L(c_i).
\end{aligned}
\tag{22}
$$

The decoding method for the code C with the decoding regions $D_1, D_2, \ldots, D_M$ is then: decode the received vector y into c_i if $y \in D_i$. In some cases, for simplicity, we can construct the decoding regions in accordance with

$$
\begin{aligned}
D_1 &= L(c_1), \\
D_2 &= L(c_2) \setminus L(c_1), \\
D_i &= L(c_i) \setminus \left(\bigcup_{j=1}^{i-1} L(c_j) \right), \quad i = 2, 3, \ldots, M.
\end{aligned}
\tag{23}
$$

This means that we decode the received vector $y \in \bigcup_{i=1}^{M} L(c_i)$ into the first c_i such that $d(y, c_i) = n$.

Since the decoding regions of (23) refer to an agreed-upon order of the codewords, the decoding is biased toward the early codewords in the list at the expense of the later ones. Such a bias is avoided by the *maximum likelihood decoding*.[6] It can be described as follows: The received vector $y \in \bigcup_{i=1}^{M} L(c_i)$ is decoded into any codeword c_i such that $Q^n(y|c_i)$ is the maximum value of $Q^n(y|c)$ over all codewords $c \in C$. If there is more than one such c_i, we choose one of them at random. It is easy to see that this is equivalent to the following decoding method: The received vector y is decoded into either one of the codewords c_i that obey $d(y, c_i) = n$, choosing one at random if there are several such c_is.

5. Decoding Error Probability

In this section, we discuss the decoding error probability and Shannon's Channel Coding Theorem for the specific channel introduced in Section 3. Some criteria for good codes for this channel are given.

For a code C with decoding regions $D_1, D_2, \ldots, D_M$, the probability e_i of the event that the vector y received by Bob is not decoded into the codeword c_i sent by Alice is given by

$$e_i = \Pr\{\mathbf{y} \notin D_i \mid \mathbf{c}_i \text{ is sent}\} = 1 - \Pr\{\mathbf{y} \in D_i \mid \mathbf{c}_i \text{ is sent}\}$$
$$= 1 - \frac{|D_i|}{3^n}, \quad i = 1, 2, \ldots, M. \tag{24}$$

The *average error probability* $\bar{e}$ is the arithmetic mean of the e_is,

$$\bar{e} = \frac{1}{M} \sum_{i=1}^{M} e_i = 1 - \frac{1}{3^n M} \sum_{i=1}^{M} |D_i|$$
$$= 1 - \frac{1}{3^n M} \left| \bigcup_{i=1}^{M} D_i \right|$$
$$= 1 - \frac{1}{3^n M} \left| \bigcup_{i=1}^{M} L(\mathbf{c}_i) \right|, \tag{25}$$

and the *maximum error probability* $e_{\max}$ is the largest one of them,

$$e_{\max} = \max_{1 \le i \le M} e_i. \tag{26}$$

Obviously, $\bar{e} \le e_{\max}$. Note that

$$\bar{e} = 0 \iff e_{\max} = 0$$
$$\iff e_i = 0 \text{ for all } i$$
$$\iff L(\mathbf{c}_i) \cap L(\mathbf{c}_j) = \emptyset, \quad i \ne j. \tag{27}$$

Hence, it follows from (19) that

$$e_{\max} \ge \bar{e} > 0. \tag{28}$$

In particular, if the decoding regions $D_1, D_2, \ldots, D_M$ are given by (23), then

$$e_1 = 1 - \frac{1}{3^n} |L(\mathbf{c}_1)| = 0,$$
$$e_i = 1 - \frac{1}{3^n} \left| L(\mathbf{c}_i) \setminus \left(\bigcup_{j=1}^{i-1} L(\mathbf{c}_j) \right) \right|$$
$$= \frac{1}{3^n} \left[|L(\mathbf{c}_i)| - \left| L(\mathbf{c}_i) \setminus \left(\bigcup_{j=1}^{i-1} L(\mathbf{c}_j) \right) \right| \right]$$
$$= \frac{1}{3^n} \left| L(\mathbf{c}_i) \cap \left(\bigcup_{j=1}^{i-1} L(\mathbf{c}_j) \right) \right|$$
$$= \frac{1}{3^n} \left| \bigcup_{j=1}^{i-1} \left(L(\mathbf{c}_i) \cap L(\mathbf{c}_j) \right) \right|, \quad i = 2, 3, \ldots, M. \tag{29}$$

The decoding error probability of a code is one of its most important performance characteristics. In this connection, we recall Shannon's Channel Coding Theorem (see Refs. 5 and 6).

Theorem 1. (*Shannon's Channel Coding Theorem*) *For any $0 < \varepsilon < 1$ and $0 < R < \log_2(4/3)$, there exists for sufficiently large n a code C of length n and size $M \doteq 2^{nR}$ such that $e_{\max} \leq \varepsilon$, and so in particular $\bar{e} \leq \varepsilon$.*

Remark 1. For fixed n and M, the best we can do is to choose a code

$$C = \{c_1, c_2, \ldots, c_M\}, \qquad c_i \in F_4^n, \tag{30}$$

such that $\left| \bigcup_{i=1}^{M} L(c_i) \right|$ is as large as possible, i.e., the average error probability $\bar{e}$ is as small as possible. For fixed n and $\bar{e} \leq \varepsilon$, in view of (25) we have to try to find a code C with the largest size M such that

$$\frac{1}{M} \left| \bigcup_{i=1}^{M} L(c_i) \right| \geq 3^n (1 - \varepsilon). \tag{31}$$

Note that for certain values of n and ε, such a code C may not exist.

6. Upper Bounds on the Decoding Error Probability

In this section, we give several upper bounds on the decoding error probability of codes for our specific channel.

Let C be a code for our specific channel. The *distance distribution* of the code C is defined by

$$A_s = \frac{1}{M} \left| \{(x, y) \in C \times C : d(x, y) = s\} \right|, \quad s = 0, 1, \ldots, n. \tag{32}$$

It is easy to see that

$$A_0 = 1, \quad \sum_{s=0}^{n} A_s = |C| = M. \tag{33}$$

Theorem 2. *Let C be a code for the specific channel in Section 3. Suppose that the distance distribution of the code C is given by $A_0, A_1, \ldots, A_n$. Then the average error probability $\bar{e}$ is upper bounded by*

$$\bar{e} \leq \frac{1}{2} \sum_{s=1}^{n} A_s \left(\frac{2}{3} \right)^s. \tag{34}$$

Proof. By the definition of $L(c_i)$ in (17) we know that if $d(c_i, c_j) = s$, then

$$\left| L(c_i) \cap L(c_j) \right| = 3^{n-s} 2^s = 3^n \left(\frac{2}{3} \right)^s, \quad s = 1, 2, \ldots, n. \tag{35}$$

Hence,

$$\sum_{1 \leq i < j \leq M} |L(\mathbf{c}_i) \cap L(\mathbf{c}_j)| = \frac{3^n M}{2} \sum_{s=1}^{n} A_s \left(\frac{2}{3}\right)^s. \tag{36}$$

By (36) and noting that $|L(\mathbf{c}_i)| = 3^n$, we obtain

$$\left| \bigcup_{i=1}^{M} L(\mathbf{c}_i) \right| \geq \sum_{i=1}^{M} |L(\mathbf{c}_i)| - \sum_{1 \leq i < j \leq M} |L(\mathbf{c}_i) \cap L(\mathbf{c}_j)|$$

$$= 3^n M - \frac{3^n M}{2} \sum_{s=1}^{n} A_s \left(\frac{2}{3}\right)^s. \tag{37}$$

Hence, (34) follows from (25) and (37). □

Theorem 3. *Let C be a code for the specific channel in Section 3. If the minimum distance $d(C) \geq d$, then the average error probability $\bar{e}$ is upper bounded by*

$$\bar{e} \leq \frac{M-1}{2} \left(\frac{2}{3}\right)^d < \frac{M}{2} \left(\frac{2}{3}\right)^d. \tag{38}$$

Proof. If the minimum distance $d(C) \geq d$, then $A_s = 0$ for $1 \leq s \leq d-1$. Hence, by Theorem 2 and (33),

$$\bar{e} \leq \frac{1}{2} \left(\frac{2}{3}\right)^d \sum_{s=d}^{n} A_s = \frac{M-1}{2} \left(\frac{2}{3}\right)^d < \frac{M}{2} \left(\frac{2}{3}\right)^d. \tag{39}$$

This completes the proof. □

Theorem 4. *Let C be a code for the specific channel in Section 3. If the minimum distance $d(C) \geq d$ and the decoding regions $D_1, D_2, \ldots, D_M$ are given by (23), then the maximum error probability $e_{\max}$ is upper bounded by*

$$e_{\max} \leq (M-1) \left(\frac{2}{3}\right)^d < M \left(\frac{2}{3}\right)^d. \tag{40}$$

Proof. If the minimum distance $d(C) \geq d$, then for $i \neq j$,

$$|L(\mathbf{c}_i) \cap L(\mathbf{c}_j)| = 3^{n-d(\mathbf{c}_i,\mathbf{c}_j)} 2^{d(\mathbf{c}_i,\mathbf{c}_j)} = 3^n \left(\frac{2}{3}\right)^{d(\mathbf{c}_i,\mathbf{c}_j)} \leq 3^n \left(\frac{2}{3}\right)^d. \tag{41}$$

It follows from (41) that for $i = 2, 3, \ldots, M$,

$$\left| \bigcup_{j=1}^{i-1} (L(\mathbf{c}_i) \cap L(\mathbf{c}_j)) \right| \leq \sum_{j=1}^{i-1} |L(\mathbf{c}_i) \cap L(\mathbf{c}_j)| \leq (i-1)3^n \left(\frac{2}{3} \right)^d$$

$$\leq (M-1)3^n \left(\frac{2}{3} \right)^d. \qquad (42)$$

Hence, by (29) and (42),

$$e_1 = 0, \quad e_i \leq (M-1) \left(\frac{2}{3} \right)^d, \quad i = 2, 3, \ldots, M. \qquad (43)$$

Therefore, $e_{\max} \leq (M-1)(2/3)^d$. $\qquad\qquad\qquad\qquad\qquad \square$

7. Linear Codes

In this section, we discuss the design of linear codes and the decoding error probability of linear codes for our specific channel. Some criteria for good linear codes are given.

First, we recall some basic concepts for linear codes. A code C over $\mathbf{F}_4$ is called a linear $[n,k]$ code over $\mathbf{F}_4$ if C is a k-dimensional subspace of $\mathbf{F}_4^n$. Note that for a linear $[n,k]$ code C over $\mathbf{F}_4$ the number of codewords is $M = 4^k$. Furthermore, we call the linear code C a linear $[n,k,d]$ code if the minimum distance of C is at least d. Let A_i be the number of codewords in C of Hamming weight i. The sequence of numbers $A_0, A_1, \ldots, A_n$ is called the *weight distribution* of C. It is well known in coding theory[7] that for a linear code C, the distance distribution of C is equal to the weight distribution of C and the minimum distance of C is equal to the minimum Hamming weight of nonzero codewords. For fixed n and k, let $d_4(n,k)$ be the maximal minimum distance of a linear $[n,k]$ code over $\mathbf{F}_4$. Tables of lower and upper bounds on $d_4(n,k)$ are available in Ref. 8.

We denote by

$$\mathcal{A} = \{ \mathbf{x} = (x_1, x_2, \ldots, x_n) \in \mathbf{F}_4^n : x_i \neq 0 \text{ for all } i \}$$
$$= \{ \mathbf{x} = (x_1, x_2, \ldots, x_n) \in \mathbf{F}_4^n : w(\mathbf{x}) = n \} \qquad (44)$$

the set of vectors with maximal Hamming weight, that is: the set of words that do not have the letter A. For a linear $[n,k]$ code C over $\mathbf{F}_4$, by (17), it is easy to check that

$$\bigcup_{\mathbf{c} \in C} L(\mathbf{c}) = \mathcal{A} + C. \qquad (45)$$

Hence, by (25), the average error probability $\bar{e}$ can be rewritten as

$$\bar{e} = 1 - \frac{1}{3^n 4^k}|\mathcal{A} + C|. \tag{46}$$

Note that $\mathcal{A} + C$ is a union of some cosets of C,

$$\mathcal{A} + C = \bigcup_{j=1}^{\alpha}(\mathbf{a}_j + C), \quad \mathbf{a}_j \in \mathcal{A}, \tag{47}$$

where $\mathbf{a}_1 + C, \mathbf{a}_2 + C, \ldots, \mathbf{a}_\alpha + C$ are some different cosets of C. This implies that $|\mathcal{A} + C| = \alpha 4^k$. Therefore, the average error probability $\bar{e}$ is also given by

$$\bar{e} = 1 - \frac{\alpha}{3^n}. \tag{48}$$

Remark 2. For fixed n and k, the best we can do is to choose a linear $[n,k]$ code C over $\mathbf{F}_4$ such that α is as large as possible. Note that even if C is optimal in this sense, the average error probability may not be small. For fixed n and $\bar{e} \leq \varepsilon$, we have to try to find a linear $[n,k]$ code C over $\mathbf{F}_4$ with the largest dimension k such that

$$\alpha \geq 3^n(1 - \varepsilon). \tag{49}$$

Note that for certain values of n and ε, such a linear code C may not exist.

Remark 3. It is known from Ref. 9 (see Problem 11 on p. 114) that for our specific channel the codes in Shannon's Channel Coding Theorem can be replaced by linear codes over $\mathbf{F}_4$, that is, for any $0 < \varepsilon < 1$ and

$$\frac{k}{n} < \frac{1}{2}\log_2(4/3) \doteq 0.2075 \tag{50}$$

there will exist, for sufficiently large n, a linear $[n,k]$ code C over $\mathbf{F}_4$ such that

$$\alpha \geq 3^n(1 - \varepsilon). \tag{51}$$

This is equivalent to the fact that $\bar{e} \leq \varepsilon$.

In general, for a linear $[n,k,d]$ code C over $\mathbf{F}_4$, by Theorem 3 we have

$$\bar{e} \leq \frac{4^k - 1}{2}\left(\frac{2}{3}\right)^d < \frac{4^k}{2}\left(\frac{2}{3}\right)^d, \tag{52}$$

and if the decoding regions $D_1, D_2, \ldots, D_{4^k}$ are given by (23), then by Theorem 4,

$$e_{\max} \leq (4^k - 1)\left(\frac{2}{3}\right)^d < 4^k\left(\frac{2}{3}\right)^d. \tag{53}$$

Furthermore, if the weight distribution $\{A_s\}_{s=0}^n$ of C is known, then by Theorem 2,

$$\bar{e} \le \frac{1}{2} \sum_{s=1}^n A_s \left(\frac{2}{3}\right)^s. \tag{54}$$

By using the Gilbert-Varshamov quasi-random construction of linear codes,[7] one can construct, for sufficiently large n, a linear $[n,k]$ code C over $\mathbf{F}_4$ with size

$$M = 4^k \doteq 4^{n[1-H_4(d/n)]} \tag{55}$$

such that the minimum distance $d(C) \ge d$, where

$$H_4(x) = x\log_4 3 - x\log_4 x - (1-x)\log_4(1-x), \quad 0 \le x \le \frac{3}{4}. \tag{56}$$

It follows from (53) that the maximum error probability $e_{\max}$ is upper bounded by

$$e_{\max} < 4^k \left(\frac{2}{3}\right)^d \doteq 4^{n[1-H_4(d/n)+(d/n)\log_4(2/3)]}. \tag{57}$$

The function of d/n in the exponent is such that

$$1 - H_4(x) + x\log_4\frac{2}{3} < 0 \Longleftrightarrow x > \beta, \tag{58}$$

where $\beta \doteq 0.4627$ is the unique solution of $1 - H_4(x) + x\log_4(2/3) = 0$. This means that, for sufficiently large n, one can construct a linear $[n,k]$ code C over $\mathbf{F}_4$ with the rate

$$\frac{k}{n} \doteq 1 - H_4(\beta) \doteq 0.1353 \tag{59}$$

such that the maximum error probability $e_{\max}$ is arbitrarily small.

8. Some Examples

In this section we give some examples of linear codes for the specific channel in Section 3 to illustrate our results. These linear codes are listed in Brouwer's tables[8] of presently best known quaternary linear codes. The value of R is obtained from $M = 2^{nR}$ in Theorem 1 and corresponds to the efficiency mentioned in Section 1.

In Table 1 we give the parameters of various codes from Ref. 8. They illustrate a general observation, namely that there is a trade-off between the simplicity of the code (short length n of the words, small number M of them) on one side and the performance of the code (large efficiency R, small average error $\bar{e}$). In addition to the codes of Table 1, we mention the following four codes of moderate length and reasonably good performance.

Example 1. Here we list explicitly known linear codes of moderate lengths.

Table 1. Examples of linear codes for the channel specified by the conditional probability distribution (7). For each code we give the number n of letters in each codeword, the dimension k of the subspace of $\mathbf{F}_4^n$, the minimum Hamming distance d, the total number M of codewords, the efficiency R, and in the last column an upper bound of (52) on the average error probability $\bar{e}$, rounded to four significant digits.
The first group on the left are six codes of length $n = 100$ with consecutive values of the dimension k. For $k \geq 16$, the upper bound on $\bar{e}$ is greater than 1, and so it is not meaningful. The second group on the left are two codes of large lengths which demonstrate that the error probability can be made very small. The group on the right are 14 codes with lengths decreasing from 50 to 10.

n	k	d	M	R	$\bar{e} \leq$	n	k	d	M	R	$\bar{e} \leq$
100	10	62	4^{10}	0.2	6.337×10^{-6}	50	5	35	1024	0.2	3.516×10^{-4}
100	11	60	4^{11}	0.22	5.704×10^{-5}	50	6	33	4096	0.24	3.165×10^{-3}
100	12	58	4^{12}	0.24	5.133×10^{-4}	48	6	32	4096	0.25	4.747×10^{-3}
100	13	56	4^{13}	0.26	4.620×10^{-3}	48	5	33	1024	0.208	7.912×10^{-4}
100	14	55	4^{14}	0.28	0.02772	47	6	31	4096	0.255	7.120×10^{-3}
100	15	52	4^{15}	0.30	0.3742	46	5	32	1024	0.217	1.187×10^{-3}
						45	5	31	1024	0.222	1.780×10^{-3}
						43	5	30	1024	0.233	2.670×10^{-3}
						42	5	29	1024	0.238	4.005×10^{-3}
200	20	109	4^{20}	0.2	3.517×10^{-8}	41	5	28	1024	0.244	6.008×10^{-3}
250	25	136	4^{25}	0.2	6.340×10^{-10}	40	4	28	256	0.2	1.502×10^{-3}
						30	3	22	64	0.2	4.277×10^{-3}
						20	2	16	16	0.2	0.01218
						10	1	10	4	0.2	0.02601

(1) A code with $n = 28$, $k = 4$, $d = 20$, $M = 256$ for which $R = \frac{2}{7} \doteq 0.2857$ and $\bar{e} < \frac{4^4}{2}(2/3)^{20} \doteq 0.03849$ is the upper bound of (52).
As is known from Ref. 8, there exists a linear $[28, 4, 20]$ code over $\mathbf{F}_4$ with the weight distribution $A_{20} = 189$, $A_{24} = 63$, $A_{28} = 3$, so that (54) gives the bound $\bar{e} \leq 0.03038$.

(2) A code with $n = 31$, $k = 4$, $d = 22$, $M = 256$ for which $R = \frac{8}{31} \doteq 0.2581$ and $\bar{e} < \frac{4^4}{2}(2/3)^{22} \doteq 0.01711$ is the upper bound of (52).
As is known from Ref. 8, there exists a linear $[31, 4, 22]$ code over $\mathbf{F}_4$ with the weight distribution $A_{22} = 141$, $A_{24} = 87$, $A_{28} = 24$, $A_{30} = 3$, so that (54) gives the bound $\bar{e} \leq 0.01216$.

(3) A code with $n = 40$, $k = 5$, $d = 28$, $M = 1024$ for which $R = \frac{1}{4} = 0.25$ and $\bar{e} < \frac{4^5}{2}(2/3)^{28} \doteq 0.006008$ is the upper bound of (52).
As is known from Ref. 8, this optimal linear code is a quasi-cyclic code. The generator matrix can be represented as $G = [G_0, G_1, G_2, G_3, G_4, G_5, G_6, G_7]$ where G_i for $0 \leq i \leq 7$ are 5×5 circulant matrices. The first row of G is given by $[10000 \quad 10120 \quad 11020 \quad 11230 \quad 12220 \quad 13130 \quad 13210 \quad 11312]$, where we identify 2 with $a \in \mathbf{F}_4$ and 3 with $b \in \mathbf{F}_4$.

where we identify 2 with $a \in \mathbf{F}_4$ and 3 with $b \in \mathbf{F}_4$.

(4) The shortened code of the previous example: $n = 39$, $k = 4$, $d = 28$, $M = 256$ for which $R = \frac{8}{39} \doteq 0.2051$ and $\bar{e} < \frac{4^4}{2}(2/3)^{28} \doteq 0.001502$.

Remark 4. For codes of small size, one can calculate the exact values of e_i, $\bar{e}$, and $e_{\max}$ by using (24)–(29). The decoding method is also computationally feasible. For codes of large size, for example $M = 4^{20}$, decoding will become an enormous computational task.

Remark 5. By using nonlinear codes, it may be possible to achieve better results on the decoding error probability, but we have not tried to search for good quaternary nonlinear codes in the literature.

Acknowledgments

This research is supported in part by the DSTA research grant R-394-000-011-422 and in part by ICITI research grant R-144-000-109-112. The work of Fang-Wei Fu is also supported by the National Natural Science Foundation of China (Grant No. 60172060), the Trans-Century Training Program Foundation for the Talents by the Education Ministry of China, and the Foundation for University Key Teacher by the Education Ministry of China.

References

1. J. M. Renes, "Spherical-code key-distribution protocols for qubits," Phys. Rev. A **70**, 052314 (2004).
2. B.-G. Englert, D. Kaszlikowski, H. K. Ng, W. K. Chua, J. Řeháček, and J. Anders, "Highly Efficient Quantum Key Distribution With Minimal State Tomography," e-print quant-ph/0412089.
3. C. H. Bennett and G. Brassard, "Quantum cryptography: Public key distribution and coin tossing," in *IEEE Conference on Computers, Systems, and Signal Processing, Bangalore, India* (IEEE, New York, 1984), p. 175.
4. D. Bruß and C. Macchiavello, "Optimal Eavesdropping in Cryptography with Three-Dimensional Quantum States," Phys. Rev. Lett. **88**, 127901 (2002).
5. T. M. Cover and J. A. Thomas, *Elements of Information Theory*. New York: Wiley, 1991.
6. R. G. Gallager, *Information Theory and Reliable Communication*. New York: Wiley, 1968.
7. F. J. MacWilliams and N. J. A. Sloane, *The Theory of Error-Correcting Codes*. Amsterdam, The Netherlands: North-Holland, 1977.
8. A. E. Brouwer, "Bounds on the minimum distance of linear codes," available at http://www.win.tue.nl/~aeb/voorlincod.html
9. I. Csiszár and J. Körner, *Information Theory: Coding Theorems for Discrete Memoryless Systems*. New York: Academic Press, 1981.

W STATE GENERATION AND EFFECT OF CAVITY PHOTONS ON THE PURIFICATION OF DOT-LIKE SINGLE QUANTUM WELL EXCITONS

C.-M. LI[1], Y.-N. CHEN[1], C.-W. LUO[1], J.-Y. HSIEH[2], D.-S. CHUU[1]

[1]*Institute and Department of Electrophysics, National Chiao Tung University, Hsinchu 30050, Taiwan.*
[2]*Department of Mechanical Engineering, Ming Hsin University of Science and Technology, Hsinchu 30401,Taiwan.*

A scheme of three-particle entanglement purification is presented in this work. The physical system undertaken for investigation is dot-like single quantum well excitons independently coupled through a single microcavity mode. The theoretical framework for the proposed scheme is based on the quantum jump approach for analyzing the progress of the trible-exciton entanglement as a series of conditional measurement has been taken on the cavity field state. We first investigate how cavity photon affects the purity of the double-exciton state and the purification efficiency in two-particle protocol. Then we extend the two-particle case and conclude that the three-exciton state can be purified into W state, which involves the one-photon-trapping phenomenon, with a high yield. Finally, an achievable setup for purification using only modest and presently feasible technologies is also proposed.

Quantum information and computation bring together the concepts from classical information science, computation theory, and quantum physics. The EPR-Bell[1] correlation, which occurs only when the quantum systems are *entangled*, generates the modern division between quantum and classical information theory. The regulation methods of quantum information processing such as quantum teleportation[2], quantum data compression[3,4], and quantum cryptography[5] rely on the transmission of maximally entangled pairs over quantum channels between a sender and a receiver. As it is well known, the quantum channels are always noisy due to inevitable interactions with environments[6]. The pairs shared by the parties may become undesired mixed states. For this reason, great attentions have been focused on the agreement of entanglement purification[7], schemes of entanglement distillation[8], and the decoherence mechanisms of quantum bits (qubits) in a reservoir[9].

Although the environment leads to the decoherence of the qubits, it may play an active role on the formation of the nonlocal effect under well considerations.

Recently, many investigations[10] have been devoted to the considerations of the reservoir-induced entanglement between two remote qubits. By utilizing the coupled field of bipartite systems[11,12,13] or, generally speaking, manipulating a third system which interacts with two remote qubits[14], many schemes have been proposed to enhance the entanglement fidelity. The scenario usually associates with a composite quantum system consisting two two-level subsystems inside a leaking optical cavity, and taking unintermitted monitor on the coupled field state. These proposals are only applicable in specific situations that the subsystems and the coupled field are in well prepared states, or in certain instant that the degree of entanglement is maximum. However, more general situations are still lacking in a entanglement distillation or entanglement generation process, especially the extension of the proposed two-particle schemes based on conditioned measurements to the protocol of multi-particle entanglement generation. This issue is crucial for both the real application of quantum communication[15] and the study on the mechanism of multi-particle entanglement.

In this paper, we present a notion of three-particle entanglement purification and an achievable setup using only modest and presently feasible technologies. The physical system undertaken for entanglement purification is dot-like single quantum well exciton coupled through a single microcavity mode. The proposed scheme is depicted in Fig.1. The whole procedure in the scheme is based on optical initialization, manipulation, and read-out of exciton state. In it, the qubit is coded in the presence of an exciton in a quantum well (QW), i.e. the exciton state in ith QW, $|e, h\rangle_i$, is considered as the logical state $|0\rangle_i$, and the vacuum state, $|0, 0\rangle_i$, which represents the state with no electron and hole, is coded as the logical state $|1\rangle_i$. The theoretical framework for the proposed scheme is based on the quantum jump approach[16]. To analyze the evolution of the double-exciton entanglement, a series of conditional measurement has been taken on the cavity field state. For accomplishing conditional measurements faithfully, the injection and leak of the cavity photon are controlled by means of the electro-optic effect. Therefore, the photon field plays an active role in the procedure of purification. According to our results of analysis, the proposed scheme can search for the decoherence-free state, i.e. the maximally entangled state, of the qubit-reservoir system in a time shorter than the characteristic time of the qubit system. Furthermore, a high purification yield and nearly unit of the entanglement fidelity can be acquired in our proposal.

First, we consider that our system consists of only two dot-like single QWs embedded inside a single-mode microcavity with the same coupling constant. We assume that the lateral size of the two QWs are sufficiently larger than the Bohr radius of excitons but smaller than the wavelength of the photon fields. Therefore, the dipole-dipole interactions and other nonlinear interactions can be neglected. The cavity mode is assumed to be resonant with the excitons. Under the rotating wave approximation, the unitary time evolution of the system is then governed by the interaction picture Hamiltonian

$$H_{2(I)} = \sum_{n=1}^{2} \hbar\gamma(a\sigma_n^+ + a^+\sigma_n^-),$$

(1)

where γ is the coupling constant, a^+ (a) is the creation (annihilation) operator of the cavity field, and σ_n^+ (σ_n) represents the creation (annihilation) operator of the excitons in the ith QW.

The dynamics of the double-dot excitons in the cavity consists of three processes: one- and two-photon absorption, one- and two-photon emission, and no photon absorption or emission. Since the formation of the maximally entangled exciton is a photon-trapping phenomenon, the main idea of the proposed scheme is to find out an eigenstate involved in the process of one-photon absorption and emission. Obviously, if the total number of quanta in the system is, m, there would be an eignstates of the Hamiltonian associates with the photon-trapping process, namely

$$|\phi\rangle = \frac{1}{\sqrt{2}}(|1\rangle_1 \otimes |0\rangle_2 - |0\rangle_1 \otimes |1\rangle_2) \otimes |m-1\rangle_c,$$

(2)

where $|m-1\rangle_c$ refers to the cavity field state with $m-1$ quanta. Once the system is in this state, the whole system does not decay at all. Accordingly, keeping the cavity mode in state $|m-1\rangle_c$, i.e. destroying the population of the cavity field state, paves the way to generate the entangled excitons

$$|\psi\rangle = \frac{1}{\sqrt{2}}(|1\rangle_1 \otimes |0\rangle_2 - |0\rangle_1 \otimes |1\rangle_2).$$

(3)

Therefore, taking a measurement on the cavity mode in order to see whether its state remains in the state or not is the key in our scheme.

For two-particle entanglement purification, the two dot-like single QWs and cavity mode is prepared in the vacuum state, $|\phi_0\rangle = |0\rangle_1 \otimes |0\rangle_2 \otimes |0\rangle_c$. A laser pulse, which is tuned to the lowest interband excitation energy, is then applied on the first QW. It can promote an electron from the valence-band to the conduction-band, hence the state, $|\phi_0\rangle = |1\rangle_1 |0_2\rangle |0\rangle_c$, prepared for purification is obtained. For the sake of generality and purpose of distillation, the state of QW can be any mixed state, ρ_ψ, except the vacuum state. Next, a pulse with m photons is injected into the microcavity. For the feasibility and the modest technology requirements, we consider here the injection of single photon, i.e. $m = 1$. The total number of quantum count of the system is two.

As the single-photon has been injected into the cavity, the total system will evolve with time, and is governed by the operator

$$U(t) =$$

$$\begin{pmatrix} 2\gamma^2 a(G-K)a^+ + 1 & -i\gamma aS & -i\gamma aS & 2\gamma^2 a(G-K)a \\ -i\gamma Sa^+ & \frac{1}{2}(\cos(\mu t) + 1) & \frac{1}{2}(\cos(\mu t) - 1) & -i\gamma Sa \\ -i\gamma Sa^+ & \frac{1}{2}(\cos(\mu t) - 1) & \frac{1}{2}(\cos(\mu t) + 1) & -i\gamma Sa \\ 2\gamma^2 a^+(G-K)a^+ & -i\gamma a^+ S & -i\gamma a^+ S & 2\gamma^2 a^+(G-K)a + 1 \end{pmatrix},$$

(4)

where $\mu^2 = K^{-1} = 2\gamma^2(2a^+a + 1)$, $G = K\cos(\mu t)$, and $S = \mu^{-1}\sin(\mu t)$. If the system evolves without interruption, it will go into a QW1- QW2-cavity field entangled state. If we take a measurement on the cavity field state at some instant, the number of the photon count of the detector may be one, two, or zero. Since the single-photon state $|1\rangle_c$ involves the photon-trapping phenomenon, we can infer that the double-QW will evolve into a maximal entangled state if the cavity mode stays in state $|1\rangle_c$ via the quantum jump approach[16]. After measuring the cavity field state, injecting a subsequent photon is necessary for the sake of keeping the photon in its state. We then let the whole system evolve for another period of time τ. Again, we proceed to measure the cavity photon to make sure whether its quantum is one or not. If the cavity photon remains in single-photon, the repetition continues; if not, the whole procedure should be started over. Therefore, after several times of successful repetitions, the state of double-exciton progresses into the state

$$\rho_{\psi_N} = (_c\langle 1|U(\tau)|1\rangle_c)^N \rho_{\psi_i} (_c\langle 1|U(-\tau)|1\rangle_c)^N / P_N,$$

(5)

where

$$P_N = \text{Tr}[(_c\langle 1|U(\tau)|1\rangle_c)^N \rho_{\psi_i} (_c\langle 1|U(-\tau)|1\rangle_c)^N],$$

(6)

is the probability of success for measuring a single-photon after N times of repetitions. The superoperator, $_c\langle 1|U(\tau)|1\rangle_c$, reveals the significant fact that it will evolve to the projection operator, $|\psi\rangle\langle\psi|$, as the successful repetitions increases. This result comes from the fact that the superoperator, $_c\langle 1|U(\tau)|1\rangle_c$, has only one eigenvalue which its absolute value equals to one[14], here, the corresponding eigenvector is just the photon-trapping state. Thus the double-exciton state will become a maximal entangled state as long as the repetitions is sufficient large. By Eq. (6), the probability of success can be evaluated

$$P_{N,1} = \frac{1}{2}(1 + \cos(\sqrt{6}\gamma\tau)^{2N}), \tag{7}$$

and the fidelity is

$$
\begin{aligned}
F_{N,1} &= \langle \psi | \rho_{\psi_N} | \psi \rangle \\
&= \frac{1}{2\cos(\sqrt{6}\gamma\tau)^{2N}}.
\end{aligned} \tag{8}
$$

Here, if the product, $\gamma\tau$, is set to be $n\pi/\sqrt{6}$, $n = 1, 2, \ldots$, then the fidelity of the double-exciton state will approach to one and the probability of success will be $1/2$ as N increases.

If the state of the cavity field is kept in $n-$photon state, rather than the state with single photon, Eq. (7) and Eq. (8) can be generalized to

$$P_{N,n} = \frac{1}{2}(1 + \cos(\sqrt{2(2n+1)}\gamma\tau)^{2N}), \tag{9}$$

and

$$F_{N,n} = \frac{1}{2\cos(\sqrt{2(2n+1)}\gamma\tau)^{2N}}. \tag{10}$$

It reveals that the number of measured cavity photon and evolution period play important roles in the trade-off between $P_{N,n}$ and $F_{N,n}$. We can choose a set of $(n, \gamma\tau)$ such that the fidelity progresses to one at the least repetitions, however, in the same time it causes the probability to reduce to a minimum. This is the usual case for purification. On the other hand, one can also find a suitable such that the probability goes to one. In this case, the system will not evolve with time and is similar to the Zeno paradox with *finite* duration between two measurements.

The concept of two-particle entanglement purification discussed above can be generalized to three-particle case directly. Here, the same approximations have been made as in the double-exciton case, and the formulation of three-exciton dynamics can be derived straightforwardly via the interaction Hamiltonian

$$H_{3(I)} = \hbar\gamma \begin{pmatrix} 0 & a & a & 0 & a & 0 & 0 & 0 \\ a^+ & 0 & 0 & a & 0 & a & 0 & 0 \\ a^+ & 0 & 0 & a & 0 & 0 & a & 0 \\ 0 & a^+ & a^+ & 0 & 0 & 0 & 0 & a \\ a^+ & 0 & 0 & 0 & 0 & a & a & 0 \\ 0 & a^+ & 0 & 0 & a^+ & 0 & 0 & a \\ 0 & 0 & a^+ & 0 & a^+ & 0 & 0 & a \\ 0 & 0 & 0 & a^+ & 0 & a^+ & a^+ & 0 \end{pmatrix}. \tag{11}$$

The superoperator that governs the progress of the three dot-like single QWs with single-photon can be worked out:

$$_c\langle 1| e^{iH_{3(I)}\tau} |1\rangle_c = g|g\rangle\langle g| + W_1|W_1\rangle\langle W_1| + T_1|T_1\rangle\langle T_1| + T_2|T_2\rangle\langle T_2|$$
$$+ e|e\rangle\langle e| + W_2|W_2\rangle\langle W_2| + T_3|T_3\rangle\langle T_3| + T_4|T_4\rangle\langle T_4|, \tag{12}$$

where g, e, W_1, W_2, T_1, T_2, T_3 and T_4 are functions of τ which correspond to the orthonormal eigenvectors

$$|g\rangle = |0\rangle_1 \otimes |0\rangle_2 \otimes |0\rangle_3,$$

$$|W_1\rangle = \frac{1}{\sqrt{3}}(|1\rangle_1 \otimes |0\rangle_2 \otimes |0\rangle_3 + |0\rangle_1 \otimes |1\rangle_2 \otimes |0\rangle_3 + |0\rangle_1 \otimes |0\rangle_2 \otimes |1\rangle_3,$$

$$|T_1\rangle = \frac{1}{\sqrt{2}}(|1\rangle_1 \otimes |0\rangle_2 \otimes |0\rangle_3 - |0\rangle_1 \otimes |0\rangle_2 \otimes |1\rangle_3),$$

$$|T_2\rangle = \frac{1}{\sqrt{6}}(|1\rangle_1 \otimes |0\rangle_2 \otimes |0\rangle_3 - 2|1\rangle_1 \otimes |0\rangle_2 \otimes |1\rangle_3 + |0\rangle_1 \otimes |0\rangle_2 \otimes |1\rangle_3),$$

$$|e\rangle = |1\rangle_1 \otimes |1\rangle_2 \otimes |1\rangle_3,$$

$$|W_2\rangle = \frac{1}{\sqrt{3}}(|0\rangle_1 \otimes |1\rangle_2 \otimes |1\rangle_3 + |1\rangle_1 \otimes |0\rangle_2 \otimes |1\rangle_3 + |1\rangle_1 \otimes |1\rangle_2 \otimes |0\rangle_3),$$

$$|T_3\rangle = \frac{1}{\sqrt{2}}(|0\rangle_1 \otimes |1\rangle_2 \otimes |1\rangle_3 - |1\rangle_1 \otimes |1\rangle_2 \otimes |0\rangle_3),$$

$$\text{and } |T_4\rangle = \frac{1}{\sqrt{6}}(|0\rangle_1 \otimes |1\rangle_2 \otimes |1\rangle_3 - 2|0\rangle_1 \otimes |1\rangle_2 \otimes |0\rangle_3 + |1\rangle_1 \otimes |1\rangle_2 \otimes |0\rangle_3). \tag{13}$$

Here $T_1 = T_2$ and $T_3 = T_4$ are two two-fold degenerate eigenvalues of the super-operator $_c\langle 1| e^{iH_{3(I)}\tau} |1\rangle_c$.

If the initial state is set equal to

$$|\phi_i\rangle = |1\rangle_1 \otimes |0\rangle_2 \otimes |0\rangle_3 \otimes |0\rangle_c, \tag{14}$$

the probability of success for finding the exciton state is in state $|W_1\rangle$ can then be worked out analytically:

$$P_N = \frac{1}{3}(\cos(\sqrt{10}\gamma\tau)^{2N} + 2\cos(\gamma\tau)^{2N}), \qquad (15)$$

and the corresponding fidelity of the three-exciton state is

$$
\begin{aligned}
F_N &= \langle W_1 | \rho_{\psi_N} | W_1 \rangle \\
&= \frac{\cos(\sqrt{10}\gamma\tau)^{2N}}{\cos(\sqrt{10}\gamma\tau)^{2N} + 2\cos(\gamma\tau)^{2N}}.
\end{aligned}
\qquad (16)
$$

Here, we can set $\gamma\tau$ to be $n\pi/\sqrt{10}$, $n = 1, 2, ...$, then the fidelity of the three-exciton state approaches to unit as N increases; meanwhile the probability of success is $1/3$. A special case, $\gamma\tau = \pi/\sqrt{6}$, is demonstrated in Fig. 2. Fig. 2(a) describes the trade-off between P_N and F_N. The purification yield $Y_N = \prod_{i=0}^{N} P_i$, which measures the fraction of surviving pairs, is also presented in Fig. 2(b).

Greenberger-Horn-Zeilinger (GHZ)[17] and W states[18,19] are two kinds of three-particle maximally entanglement. The former involves three-photon trapping phenomenon which can be described through the state vector

$$|\psi_{GHZ}\rangle = \frac{1}{\sqrt{3}}(|0\rangle_1 \otimes |0\rangle_2 \otimes |0\rangle_3 - |1\rangle_1 \otimes |1\rangle_2 \otimes |1\rangle_3). \qquad (17)$$

and W state is a one-photon trapping state

$$|\psi_W\rangle = \frac{1}{\sqrt{3}}(|1\rangle_1 \otimes |0\rangle_2 \otimes |0\rangle_3 + |0\rangle_1 \otimes |1\rangle_2 \otimes |0\rangle_3 + |0\rangle_1 \otimes |0\rangle_2 \otimes |1\rangle_3). \quad (18)$$

Although these states associated with different physical phenomenon, they possess the same degree of entanglement[18,19]. In our scheme, any initial mixed state of the three-exciton state can be purified into W state except the vacuum. However, the symmetry properties of qubit-environment interactions for GHZ state is quite different from three-exciton W state except the vacuum, thus it can not be generated via the approach of conditional measurement.

The whole concepts for the experiments are schematically shown in Fig. 3. A Ti:sapphire laser supplies the pulse light source for all of the operation photons in our devices such as resonant photon, 2ω photon, and pulse E field. The timing between all of the operation photons could be controlled precisely by delay stages. Also, the detection of photodiodes could be precisely triggered by laser pulses. Firstly, one of the dot-like single quantum wells is excited by the $3eV$ photon from 2ω generator performed through a nonlinear crystal, e.g. BBO or LBO crystal. Then, a resonant photon with vertical linear polarization generated via a quartz plate is injected into the cavity which is constructed by the ZnTe with both Au films. Meanwhile, through the pulse **E** field[15] with appropriate magnitude[18] and pulse width of ˜3.3*ps* as shown in Fig. 4, the linear polarization of injected photon is rotated from vertical to horizontal via the Electro-optic effect in ZnTe[19]. After the sufficient evolution time (T_{ev} ˜20*ps*, the time period, τ, between two measurements) with dot-like single quantum well excitons, the photon in cavity can be leaked out the cavity by a pulse **E** field with suitable timing and detected by a single photon avalanche diode (SPAD, detector 2 in Fig. 3). This procedure would be repeated until finishing the purification. During this procedure, the photo luminescence lifetime of dot-like single quantum well exciton can be measured by the other SPAD as shown in Fig. 4 (detector 1 in Fig. 3).

To summarize, a three-particle entanglement purification scheme based on conditioned measurements has been proposed in this work. We investigate the entanglement generation of dot-like single quantum well excitons coupled through a single microcavity mode. As shown in Eq. (9) and Eq. (10), we first consider how the cavity photon affects the purity of the exciton state and the purification efficiency in two-particle protocol. The trade-off between $P_{N,n}$ and $F_{N,n}$ reveals whether the double excitons can be purified into a photon trapping, decoherence-free state efficiently depending on the number of photon counts in the repeated measurements. Moreover, we conclude that the three-exciton state can be purified into W state with a well yield but smaller than that in two-particle case. Finally, a feasible experimental setup for optical initialization, manipulation, and read-out of exciton state is also presented.

This work is supported partially by the National Science Council, Taiwan under the grant numbers NSC 93-2112-M-009-037 and NSC 94-2120-M-009-002.

References

1. A. Einstein, B. Podolsky, and N. Rosen, Phys. Rev. 47, 777 (1935); J. S. Bell, Rev. Mod. Phys. 38, 447 (1966).
2. C. H. Bennett, G. Brassard, C. Crepeau, R. Jozsa, A. Peres, and W. K. Wootters, Phys. Rev. Lett., 70, 1895 (1993).
3. B. Schumacher, Phys. Rev. A, 51, 2738 (1995).
4. R. Jozsa and B. Schumacher, J. Modern Optics, 41, 2343 (1994).
5. A. K. Ekert, Phys. Rev. Lett., 67, 661 (1991).
6. W. H. Zuerk, Mod. Phys. 75, 715 (2003).
7. C. H. Bennett, G. Brassard, S. Popescu, B. Schumacher, J. A. Smolin, and W. K. Wootters, Phys. Rev. Lett., 76, 722 (1996); D. Deutsch, A. Ekert, R. Jozsa, C. Macchiavello, S. Popescu, and A. Sanpera, Phys. Rev. Lett., 77, 2818 (1996); J.-Y. Hsieh, C.-M. Li, and D.-S. Chuu, Physics Letters A, 328, 94 (2004).

8. L.-M. Duan, M. D. Lukin, J. I. Cirac, and P. Zoller, Science, 414, 413 (2001).
9. S. Bose, I. Fuentes-Guridi, P. L. Knight, and V. Vedral, Phys. Rev. Lett. 87, 050401 (2001); 87, 279901(E) (2001); R. W. Rendell and A. K. Rajagopal, Phys. Rev. A 67, 062110 (2003).
10. Daniel Braun, Phys. Rev. Lett. 89, 277901 (2002); Y. N. Chen, D. S. Chuu, and T. Brandes, Phys. Rev. Lett. 90, 166802 (2003) ; M. Paternostro, W. Son, and M. S. Kim, Phys. Rev. Lett. 92, 197901 (2004).
11. X. X. Yi, C. S. Yu, L. Zhou, and H. S. Song, Phys. Rev. A 68, 052304 (2003).
12. C. Cabrillo, J.I. Cirac, P. Garcia-Fernandez, and P. Zoller, Phys. Rev. A 59, 1025 (1999)
13. M. Plenio, S.F. Huelga, A. Beige, and P.L. Knight, Phys. Rev. A 59, 2468 (1999).
14. H. Nakazato, T. Takazawa, and K. Yuasa, Phys. Rev. Lett. 90, 060401 (2003); H. Nakazato, M. Unoki, and K. Yuasa, Phys. Rev. A 70, 012303 (2004); G. Compagno, A. Messina, H. Nakazato, A. Napoli, M. Unoki, and K. Yuasa, Report No. quant-ph/0405074.; L.-A. Wu, D. A. Lidar, S. Schneider, Phys. Rev. A 70, 032322 (2004).
15. J. Joo, Y. Park, S. Oh and J. Kim, New Journal of Physics 5, 136 (2003).
16. M. Plenio and P.L. Knight,, Rev. Mod. Phys. 70, 101 (1998).
17. D. M. Greenberger, M. A. Horne, and A. Zeilinger, in Bell's Theorem, Quantum Theory, and Conceptions of the Universe, edited by M. Kafatos (Kluwer, Dordrecht, 1989), p. 69.
18. W. Dur, G. Vidal, and J. I. Cirac, Phys. Rev. A 62, 062314 (2000).
19. A. Acin, D. Brub, M. Lewenstein, and A. Sanpera, Phys. Rev. Lett. 87, 040401 (2001).
20. Actually, the pulse E field is the "THz radiation" which could be easily generated from GaAs semiconductor or several nonlinear crystal such as GaSe [21, 22].
21. K. Reimann, et al., Opt. Lett. 28, 471 (2003).
22. C. W. Luo, et al., Semicond. Sci. Technol. 19, S285 (2004).
23. According to the electro-optic modulation theory, the half-wave voltage derived from phase retardation between principal axes is given by $V\pi = (d\lambda)/(L2n^3 r_{41})$. For our device, where $d = 2\mu m$ is the thickness of ZnTe and is along the pulse E field, $L = 1$mm is the length of ZnTe, i.e. the optical path length of injected photon,$\lambda = 713 nm$ is the wavelength of injected photon, $n = 2.906$ is the index of refraction for 713 nm, $r_{41} = 3.9^{-12} m/V$ is the electro-optic coefficient. Then we can evaluate that$V\pi = 7.45V$. We need ~7.45V to rotate the polarization of injected photon from vertical to horizontal. This means that the magnitude of pulse E field with the ZnTe of 2μm thickness is ~$37.25KV/cm$.
24. Q. Wu and X.-C. Zhang, Appl. Phys. Lett. 71, 1285 (1997).

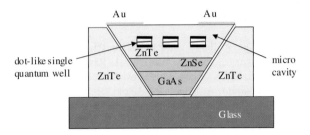

Figure 1.   The quantum devices with triple dot-like quantum wells inlaid in a microcavity which is constructed by a ZnTe medium and two Au mirrors, will be prepared by the MBE, the e-beam lithography, and the conventional semiconductor processing.

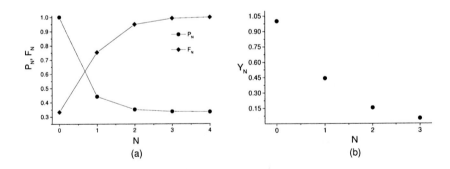

Figure 2.   (a) The probability of success for measuring the same photon state, P_N, and the fidelity of the exciton state, F_N, and (b) the corresponding yield, Y_N, for W state generation.

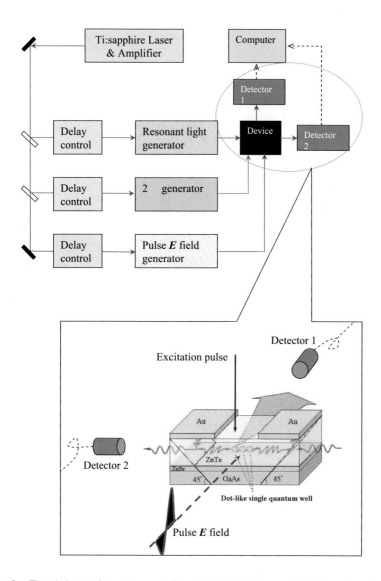

Figure 3. The whole experimental concepts for optical initialization, manipulation, and read-out of exciton state.

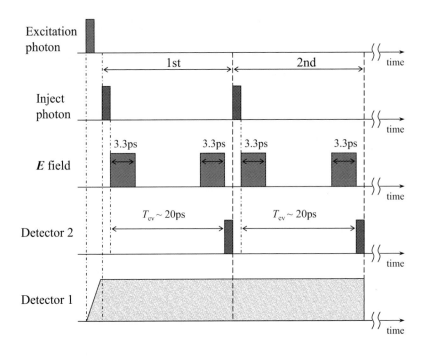

Figure 4.   The flowchart for W state generation. The inset quantum device is the same with the device in Figure 1.

A UNIVERSAL QUANTUM ESTIMATOR

L. C. KWEK[1,2], K. W. CHOO[2], JIANGFENG DU[1], ARTUR K. EKERT [1,3], CAROLINA MOURA ALVES [4], MICHAł HORODECKI[5], PAWEł HORODECKI[6], D. KASZLIKOWSKI[1], NAVEEN NAZIMUDEEN[7], C.H. OH [1], AND DANIEL K. L. OI[3]

[1] *Department of Physics, National University of Singapore, 2 Science Drive 3, Singapore 117542;*
[2] *National Institute of Education, Nanyang Technological University, 1 Nanyang Walk, Singapore 639798;*
[3] *Centre for Quantum Computation, DAMTP, University of Cambridge, Cambridge CB3 0WA, UK;*
[4] *Centre for Quantum Computation, Clarendon Laboratory, University of Oxford, Parks Road, Oxford OX1 3PU, U.K.;*
[5] *Institute of Theoretical Physics and Astrophysics, University of Gdańsk, 80-952 Gdańsk, Poland;*
[6] *Faculty of Applied Physics and Mathematics, Technical University of Gdańsk, 80-952 Gdańsk, Poland;*
[7] *Electrical and Electronics Engineering, Nanyang Technological University, Singapore*

Almost all computational tasks in the modern computer can be designed from basic building blocks. These building blocks provide a powerful and efficient language for describing algorithms. In quantum computers, the basic building blocks are the quantum gates. In this tutorial, we will look at quantum gates that act on one and two qubits and briefly discuss how these gates can be used in quantum networks.

1. Introduction

There has been much progress in quantum information processing in the last few years. However, the possibility of building a sizeable quantum information processor based on current technology remains an open question. As an intermediate goal, there has been a number of recent proposals for a fixed architecture or network [1]which does not perform arbitrary computation but could extract important properties of a quantum state ρ, such as its purity, degree of entanglement, or its spectrum which are also important in quantum information science.

Properties of quantum states can generally be quantified in terms of linear or non-linear functionals of ρ. Linear functionals, such as average values of observables $\{A\}$, which given by Tr$A\rho$, are usually needed since they correspond to directly measurable quantities. Non-linear functionals of state, such as the von Neumann entropy $-\text{Tr}\rho \ln\rho$, determination of eigenvalues, or a measure of purity $\text{Tr}\rho^2$, are usually extracted from ρ by conventional tomography, namely, ρ is first estimated and once a sufficiently precise classical description of ρ is available, classical evaluations of the required functionals can be made.

If there is only a limited supply of physical objects in state ρ, a direct estimation of a specific quantity would often be more efficient and more desirable [2]. Moreover, we learnt from our study of quantum algorithm that it is often not necessary to have a complete reconstruction of a function for a particular estimation or computation. For example, in the Deutsch algorithm, one does not need to compute the function values for all inputs in order to determine whether a function is constant or balanced. The estimation of purity of ρ also does not require knowledge of all matrix elements of ρ, thus any prior state estimation procedure followed by classical calculations is, in this case, inefficient. In general, one can resort to a fixed architecture to avoid tomography and faclitate the estimate nonlinear functionals of ρ more directly quantum networks [3,4].

The quantum network [1] analyzed here is a simple but fairly general purpose quantum network based on two Hadamard gates(H), a phase gate(Φ) and a controlled-$U(\widetilde{U})$ gate. The network is shown in figure 1. In this paper, we proceed to analyze this quantum network in some detail, adopting a pedagogical approach in the presentation.

To begin, let us write down the matrices for the various gates in this quantum network. These gates are the Hadamard and phase gates acting on single qubit gates and the controlled-U gate acting on two or more qubits:

$$H \equiv \frac{1}{\sqrt{2}}\begin{pmatrix} 1 & 1 \\ 1 & -1 \end{pmatrix} \qquad \Phi \equiv \begin{pmatrix} 1 & 0 \\ 0 & e^{i\phi} \end{pmatrix} \qquad \widetilde{U} \equiv \begin{pmatrix} I & 0 \\ 0 & U \end{pmatrix} \qquad (1)$$

The matrix $\widetilde{U}$ that describes controlled-U above is a block matrix. We can consider the input state to the quantum network as a trivial tensor product of a single qubit in the state($|0\rangle$) (System S_A) with density operator ρ (System S_B), specifically as $|0\rangle\langle 0| \otimes \rho_B$. Let us study state after the various operations of the intermediate gates. The Hadamard and phase operations are first performed on the first qubit. The final output is given by

$$\rho_{out} = (H \otimes I)(U(\rho' \otimes \rho_B)U^\dagger)(H \otimes I)$$
$$= \frac{1}{4}\begin{bmatrix} \rho_B + U\rho_B U^\dagger + e^{i\phi}U\rho_B + e^{-i\phi}\rho_B U^\dagger & \rho_B - U\rho_B U^\dagger - e^{-i\phi}\rho_B U^\dagger + e^{i\phi}U\rho_B \\ \rho_B - U\rho_B U^\dagger - e^{i\phi}U\rho_B + e^{-i\phi}\rho_B U^\dagger & \rho_B + U\rho_B U^\dagger - e^{-i\phi}\rho_B U^\dagger - e^{i\phi}U\rho_B \end{bmatrix}$$

To obtain the density operator of subsystem A one takes a partial trace of the over

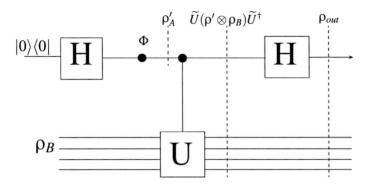

Figure 1.   A general quantum network

system B giving

$$\rho_A^{out} \text{Tr}_B(\rho_{out}) =$$

$$\frac{1}{4}\begin{bmatrix} \text{Tr}(\rho_B + U\rho_B U^\dagger + e^{i\phi}U\rho_B + e^{-i\phi}\rho_B U^\dagger) & \text{Tr}(\rho_B - U\rho_B U^\dagger - e^{-i\phi}\rho_B U^\dagger + e^{i\phi}U\rho_B) \\ \text{Tr}(\rho_B - U\rho_B U^\dagger - e^{i\phi}U\rho_B + e^{-i\phi}\rho_B U^\dagger) & \text{Tr}(\rho_B + U\rho_B U^\dagger - e^{-i\phi}\rho_B U^\dagger - e^{i\phi}U\rho_B) \end{bmatrix}$$

The diagonal entries of ρ_A^{out} give us the probability of observing the system in state $|0\rangle$ or $|1\rangle$. In particular if we obtain the probability of observing $|0\rangle$, P_0, as

$$P_0 = \frac{1}{4}\text{Tr}(\rho_B + U\rho_B U^\dagger + e^{i\phi}U\rho_B + e^{-i\phi}\rho_B U^\dagger)$$

$$= \frac{1}{4}(2 + e^{i\phi}\text{Tr}(U\rho_B) + e^{-i\phi}\text{Tr}(\rho_B U^\dagger))$$

$\text{Tr}(\rho_B U)$ being a complex number can be written as $|\text{Tr}(\rho_B U)|e^{i\alpha}$

$$P_0 = \frac{1}{4}(2 + e^{i\phi}|\text{Tr}(\rho_B U)|e^{i\alpha} + e^{-i\phi}|\text{Tr}(\rho_B U)|e^{-i\alpha})$$

$$= \frac{1}{4}(2 + |\text{Tr}(\rho_B U)|(e^{i(\phi+\alpha)} + e^{-i(\phi+\alpha)}))$$

$$= \frac{1}{4}(2 + 2|\text{Tr}(\rho_B U)|\cos(\phi+\alpha))$$

$$= \frac{1}{2}(1 + |\text{Tr}(\rho_B U)|\cos(\phi+\alpha))$$

By varying the phase ϕ of the phase gate, one sees a variation in the probability of P_0 at the output. We can then define visibility as the difference between the maximum and minimum values in the variation of the probability of observing the state $|0\rangle$ when one varies ϕ.

$$\text{visibility} = \frac{1}{2}(1 + |\text{Tr}(\rho_B U)|) - \frac{1}{2}(1 - |\text{Tr}(\rho_B U)|)$$

$$= |\text{Tr}(\rho_B U)| \tag{2}$$

It is interesting to note that Eq(2) provides an estimate of the average value of the unitary operator U in state ρ.

2. Estimation of quantum state

The basic idea in state estimation is to make use of the basis set of the density matrix. specifically, the expression in Eq(2) allows us to estimate an unknown quantum state ρ as long as we can estimate $\text{Tr}U_k\rho$ for a set of unitary operators U_k which form a basis in the vector space of density operators. . Note that quantum state estimation could also be done with various other schemes of quantum tomography. However, this network provides a fixed architecture for the purpose.

As an example, one can consider the basis matrix set of a n-level density matrix ρ as the set of matrices

$$\{\mathbb{I}, \lambda_1, \lambda_2, \ldots, \lambda_{2^n-1}\} \tag{3}$$

such that $\text{Tr}(\lambda_i) = 0$ and $\text{Tr}(\lambda_i \lambda_j) = 2\delta_{ij}$. Any state ρ can be written as a linear combination of the base matrices

$$\rho = \frac{1}{N}\left[1 + \sqrt{\frac{n(n-1)}{2}}\vec{m}.\vec{\lambda}\right] \tag{4}$$

Figure 1 shows the quantum network, where S_B is the quantum state with unknown density matrix ρ and the controlled-U operation is the elements of the basis set. The experiment is performed $2^n - 1$ times, each time using the next element of the basis set as controlled-U. With each experiment, we would get an element of the vector $\vec{m}$.

$$\text{Tr}(\rho\lambda_i) = \text{Tr}\left[\frac{1}{n}\left[\lambda_i + \sqrt{\frac{n(n-1)}{2}}\sum_j m_j\lambda_i\lambda_j\right]\right]$$

$$= \frac{\text{Tr}(\lambda_i)}{n} + \sqrt{\frac{(n-1)}{2n}}\sum_j m_j\text{Tr}(\lambda_i\lambda_j)$$

$$= \sqrt{\frac{(n-1)}{2n}}\sum_j m_j 2\delta_{ij}$$

$$= \sqrt{\frac{2(n-1)}{n}}m_i \tag{5}$$

As a further example, we can also consider the specific case of a 2×2 density matrix, where the basis is $\{\mathbb{I}, \sigma_1, \sigma_2, \sigma_3\}$ known as the Pauli matrices. An arbitrary 2×2 density matrix can be represented as $\frac{1}{2}(1 + n_1\sigma_1 + n_2\sigma_2 + n_3\sigma_3)$. The following shows the case where the controlled-U is the first Pauli matrix.

$$\mathrm{Tr}(\rho\sigma_1) = \mathrm{Tr}(\frac{1}{2}(\sigma_1 + n_1 + n_2\sigma_2\sigma_1 + n_3\sigma_3\sigma_1)$$

$$= n_1 + \frac{n_2}{2}\mathrm{Tr}(\sigma_2\sigma_1) + \frac{n_3}{2}\mathrm{Tr}(\sigma_3\sigma_1)$$

$$= n_1$$

Repeating the experiment with σ_2 and σ_3 as controlled-U would give n_2 and n_3 respectively from which ρ can be constructed.

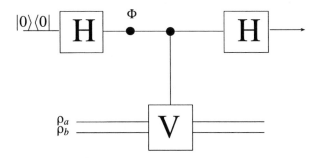

Figure 2.   Quantum Network with Swap operator

3. Quantum Network with swap operator

It is possible to modify minimally the quantum network by changing the controlled-U to be controlled-V where V is the swap operator. The swap operator is defined as $V|\alpha\rangle_A|\beta\rangle_B = |\beta\rangle_A|\alpha\rangle_B$. To this end, we examine a case in which S_B is a quantum state of two separable subsystems. The density matrix of the composite system is $\rho = \rho_1 \otimes \rho_2$, where ρ_1 and ρ_2 are the density operators of each subsystem.

$$v = \mathrm{Tr}(\rho_1 \otimes \rho_2)V = \mathrm{Tr}\rho_1\rho_2 \tag{6}$$

From section 1 we know that for any controlled-U the output visibility would be $\mathrm{Tr}(\rho U)$. Since the controlled-U is replaced by a controlled-swap the output visibility would be $\mathrm{Tr}(\rho V)$:

$$(\rho_1 \otimes \rho_2)V = \sum_{ijkl} P_{ij}P'_{kl}|ik\rangle\langle lj|$$

$$\mathrm{Tr}((\rho_1 \otimes \rho_2)V) = \sum_{rs} P_{rs}P'_{sr}$$

On the other hand,

$$\mathrm{Tr}(\rho_1\rho_2) = \sum_{jm} P_{mj}P'_{jm}$$

Thus replacing controlled-U with controlled-V modifies the output visibility expression from $\mathrm{Tr}(\rho_1 \otimes \rho_2 U)$ to $\mathrm{Tr}(\rho_1\rho_2)$. It is in fact possible to extend this result for S_B as a composite system of an arbitrary n subsystems. The result for n subsystems may be written as.

$$\mathrm{Tr}((\rho_1 \otimes \rho_2 \dots \otimes \rho_n)V^{(n)}) = \mathrm{Tr}(\rho_1\rho_2\dots\rho_n) \qquad (7)$$

The controlled-U for the general n case would be a controlled shift operation which is a generalisation of the controlled-SWAP gate. Shift operator for n-dimensional state can be defined as

$$V^{(n)}|\phi_1\rangle|\phi_2\rangle\dots|\phi_n\rangle = |\phi_n\rangle|\phi_1\rangle\dots|\phi_{n-1}\rangle \qquad (8)$$

which in the outer product notation can be written as

$$\sum_{a_i} |a_n a_1 \dots a_{n-1}\rangle\langle a_1 a_2 \dots a_n| \qquad (9)$$

where $|a_1 a_2 \dots a_n\rangle$ forms an orthogonal basis.

This modified quantum network can be employed further in many other quantum experiments and it has some very useful applications. In the sections that follow, we trace some of these applications in some details.

4. Orthogonality of states

The quantum network shown in figure 4 can be used to detect the amount of orthogonality of unknown pure states. This means that the fixed architecture can

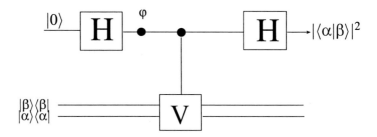

Figure 3.   Quantum network for orthogonality measurement

provide a measure of distinguishability between two states $|\alpha\rangle$ and $|\beta\rangle$. Indeed,

$$
\begin{aligned}
\mathrm{Tr}(\rho_a\rho_b) &= \mathrm{Tr}(|\alpha\rangle\langle\alpha|\beta\rangle\langle\beta|) \\
&= \sum_i \langle i|\alpha\rangle\langle\alpha|\beta\rangle\langle\beta|i\rangle \\
&= \sum_i \langle\alpha|\beta\rangle\langle\beta|i\rangle\langle i|\alpha\rangle \\
&= \langle\alpha|\beta\rangle\langle\beta|\alpha\rangle \\
&= \langle\alpha|\beta\rangle\langle\alpha|\beta\rangle^* \\
&= |\langle\alpha|\beta\rangle|^2
\end{aligned}
\tag{10}
$$

5. Spectrum of density matrix

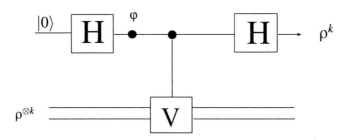

Figure 4.   Quantum network for finding density matrix spectrum

The ability to determine the eigenvalues and eigenvectors of a state can provide important data regarding the state. From such measurements, we can understand important properties of the state, such as separability and entanglement or purity of a state etc. In general for $d \times d$ density matrix, there are d eigenvalues, therefore one requires d independent equations are required to solve for them.

Using the quantum network shown in figure 4, one can perform the experiment $d-1$ times, each time with $\rho^{\otimes k}$ as S_B, where k varies from 2 to d. The output visibilities of each experiment would give $\text{Tr}(\rho^k$ as the output. In this way,we get $d-1$ independent equations which can be solved to find their value.

6. Extremal Eigenvalues

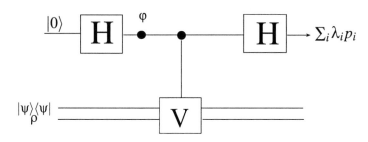

Figure 5.   Quantum Network to find extremal eigenvalues

In the previous section we had presented a quantum network to find the entire spectrum of a density. Here we use the same network to determine the extremal eigenvalues and eigenvectors of a system having density matrix ρ. Extremal eigenvalues can provide information regarding the amount of entanglement of a bipartite system.

Figure 5 shows how the quantum network can be used in such experiment. S_B is prepared in a state $|\psi\rangle\langle\psi| \otimes \rho$, where $|\psi\rangle$ is a state appropriately chosen.

Any density operator can be written as $\sum_i \lambda_i |\eta_i\rangle\langle\eta_i|$ where λ_i are the eigenvalues and $|\psi\rangle$ are the corresponding eigenvalues of the system. This follows from the spectral decomposition theorem. Visibility at the output of this quantum network will then be

$$v = \text{Tr}\big(|\psi\rangle\langle\psi|\rho\big)$$
$$= \text{Tr}\big(|\psi\rangle\langle\psi| \sum_i \lambda_i |\eta_i\rangle\langle\eta_i|\big)$$
$$= \sum_i \lambda_i |\langle\psi|\eta_i\rangle|^2 = \sum_i \lambda_i p_i$$

Since $\sum p_i = 1$, $\sum_i \lambda_i p_i$ is a convex sum of eigenvalues λ_i of ρ. If we can now change the value of p_i by varying the ψ. The visibility would be maximum when $|\psi\rangle = |\eta_{max}\rangle$, corresponding to eigenvector with the maximum eigenvalue.Thus to find the maximum (minimum) value, the network is repeatedly scanned with different ψ.

7. Experimental Implementation

Recently, we implemented the simple network using the three-qubit alanine(in D_2O) NMR QIP through the reading pulse sequence. Spin-C_1 serves as qubit-1, Spin-C_2 serves as qubit-2, which holds $|\psi_1\rangle$, and Spin-C_3 serves as qubit-3, which holds $|\psi_2\rangle$. Hydrogen nuclei are decoupled during the experiments.

The pseudo-pure state was implemented by spatial label using the method described in Ref. [5], but the swap gate was implemented by the pulse sequence described in Ref. [6] instead of three CNOT gates for simplifications. The pulse sequence for the scheme is shown in Fig. 6. The main pulse sequence includes three parts: 1. prepare $|\psi_1\rangle = \cos\frac{\theta_1}{2}|0\rangle - e^{i\varphi_1}\sin\frac{\theta_1}{2}|1\rangle$ and $|\psi_2\rangle = \cos\frac{\theta_2}{2}|0\rangle - e^{i\varphi_2}\sin\frac{\theta_2}{2}|1\rangle$, this could be implemented by selective pulses on Spin-C_2 and Spin-C_3. 2. The network is given in Fig. 7. The two Hadamard gates were implemented by $R_\varphi^1(90)$ and $R_{-y}^1(90)$. The φ gate was reduced, it was put on the phase of the first Hadamard gate. the Fredkin gate is implemented with three transition pulses TP1,TP2,TP3 [7,8]. 3. Reading pulses. The way that it works is explained below.

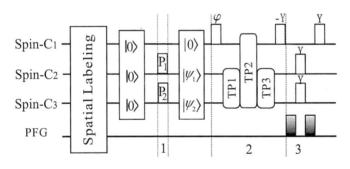

Figure 6

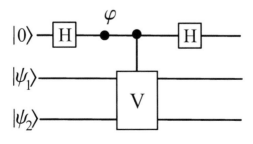

Figure 7

After the first gradient pulse the spin system becomes

$$\sum_{i=1}^{3} a_i I_z^i + a_{12}I_z^1 I_z^2 + a_{13}I_z^1 I_z^3 + a_{32}I_z^3 I_z^2 + a_{123}I_z^1 I_z^2 I_z^3, \tag{11}$$

apply $R_y^2(90)$ and $R_y^3(90)$ after it, the spin system becomes

$$a_1 I_z^1 + a_{12}I_z^1 I_x^2 + a_{13}I_z^1 I_x^3 + a_{123}I_z^1 I_x^2 I_x^3 + \sum_{i=2}^{3} a_i I_x^i + a_{32}I_x^3 I_x^2, \tag{12}$$

after the second gradient pulse the spin system becomes

$$a_1 I_z^1. \tag{13}$$

When all the the reading pulses were applied the integral area of the spectra from the state $a_1 I_x^1$ is proportional to the difference of the two probabilities $(\rho_{00} - \rho_{11})$ – the overlap of $|\psi_1\rangle$ and $|\psi_2\rangle$. Processing of tracing out Spin-C_2 and Spin-C_3 was done by integrating the entire multiplet of the Spin-C_1. The results of the NMR experiments confirmed and agrees with the theorectical calculations and proposals [9].

8. Discussion

This brief tutorial provides some possible applications of a fixed architecture which can be used for the estimation of various linear and non-linear functionals. For instructional purposes, we have not included the use of the network for estimating quantum channels [1]or remote structural physical approximation [10].

References

1. A. K. Ekert, C. Moura Alves, D. K. L Oi, M. Horodecki, P. Horodecki, L. C. Kwek, *Phys. Rev. Lett.* **88**, 217901 (2002)
2. P. Horodecki, A. Ekert, quant-ph/0111064
3. D. Deutsch, *Proc. R. Soc. Lond. A* **425** 73 (1989).
4. A. Barenco, C. H. Bennett, R. Cleve, D. P. DiVincenzo, N. Margolus, P. W. Shor, T. Sleator, J. Smolin and H.Weinfurter, *Phys. Rev. A* **52** 3457 (1995).
5. U. Sakaguchi, H. Ozawa, and T. Fukumi, *Phys. Rev. A* **61**, 042313 (2000).
6. Z.L. Madi, R. Bruschweiler, and R.R. Ernst, *J Chem. Phys.* **109**, No.24, 10603 (1998).
7. Xue Fei, Shi MingJun, Du JiangFeng, *et al.*, *Chin. Phys. Lett.* **19** 1048 (2002).
8. Du JiangFeng, Shi MingJun, Wu JiHui, *et al.*, *Phys. Rev. A* **63**, 042302 (2001).
9. Results have been submitted for publication elsewhere.
10. C. M. Alves, P. Horodecki, D. K. L. Oi, L. C. Kwek, and A. K. Ekert, *Phys. Rev. A* **68**, 032306 (2003)

TOWARD A PRACTICAL ENVIRONMENT FOR QUANTUM PROGRAMMING*

S. YAMASHITA, M. NAKANISHI, AND K. WATANABE

Graduate School of Information Science,
Nara Institute of Science and Technology
8916-5 Takayama, Ikoma, Nara 630-0192, JAPAN
E-mail: {ger, m-naka, watanabe}@is.naist.jp

This paper proposes a practical framework for quantum programming. In our framework, the parts of a program to be performed on a quantum computer are almost automatically determined, and the other parts are performed on a classical computer. We only consider Grover Search to be performed on a quantum computer in the framework because the other quantum algorithms known so far cannot be applied to general cases. By considering only Grover Search, we have several advantages that show our framework is really practical.

1. Introduction

Many efforts have focused on the physical realization of quantum computers and new efficient quantum algorithms for the realization of quantum computation. Such researches are obviously indispensable to perform quantum computation. However, even after the success of these researches, we cannot obtain a better computing environment than today's from a *practical* viewpoint because at least two of the following issues might exist and be obstacles to realize a *practical* quantum computing environment.

The first issue is that for some computational tasks, the cost of quantum computation (i.e., the time and money to perform computation) is much higher than classical computation. The number of necessary computational steps (in the Turing machine model) has been proven to be the same for some computational tasks between quantum and classical cases. For such computational tasks, quantum computers may not be a good choice since the number of necessary primitive operations to realize one computational step is much larger in quantum computation than in classical. It can also be expected that the cost of one primitive operation

---

*This work is supported by the Japanese Ministry of Internal Affairs and Communications under SCOPE project.

is higher than in classical cases. In other words, for some tasks, we should use a classical instead of a quantum computer.

The second issue is that it is almost impossible to manually design a desired sequence of primitive operations (i.e., a quantum circuit) for problems of practical sizes. Therefore, as in the classical case, we need both an efficient way of describing an algorithm in higher levels (i.e., programming language) and an automatic way of transforming it to primitive operations.

These two issues, which have not been discussed seriously in the research community, are the obstacles for practical quantum computation. Note that it is not trivial to deal with the above issues by simply using the techniques currently available in classical programming environments.

Therefore, in this paper, we propose a framework to initiate research that deals with the above issues. In the framework, a programmer can write a program by using (exactly the same) C++ language whereas most existing researches demand some extra knowledge to write a program for quantum computers[1,2,3]. Then the framework automatically extracts the parts of the program that can be performed faster by Grover Search[4] and generates corresponding quantum circuits and interface codes for a classical computer. We can also simulate the entire program in only a classical programming environment since we can easily prepare C++ source codes to simulate quantum circuits.

2. Overview of Proposed Strategy

We present an overview of the proposed framework in Section 2.3 after explaining the two main features of our strategy in Secs. 2.1 and 2.2.

2.1. *Classical and Quantum Co-design*

Let us start with an observation on quantum computations from the view point of *real* computation time, i.e., not asymptotic time complexity as usually discussed in this community. It is well-known that a quantum computer can perform all kinds of classical computation with only polynomial overhead. The overhead is mainly due to the simulation of classical computation by only using quantum elementary gates, i.e., reversible logic gates. It should also be assumed that one logical operation on a quantum computer is much slower than on a classical computer. Therefore, from a practical viewpoint, it may not be wise to use a quantum computer to only simulate classical computations: a quantum computer should be used only for parts that can actually receive the quantum speed up.

Therefore, we consider a quantum computer as a coprocessor that can boost the speed of some (not all) parts of a program on a classical computer. A similar computation model called "QRAM" has been discussed in the literature[1].

We also consider that the separation of quantum and classical parts should be done automatically by our programming framework as much as possible. We call such separation **classical and quantum co-design** borrowing the terminology "hardware-software co-design[5]." If we consider the *practical* usage of quantum computers, the issue of classical and quantum co-design becomes important, and thus we propose our framework as a promising classical and quantum co-design methodology.

2.2. *Limitations to Grover Search*

As we mentioned, our classical and quantum co-design framework should determine a part in a program for quantum computation. Then the first question is: what types of parts are candidates for quantum computation? Many algorithms show quantum speed up, e.g., Shor's factoring algorithm[6], Grover Search[4], Quantum Walk, etc. Among them, only Grove Search has potential for general purposes. In other words, other quantum algorithms are useful for very specific situations, and thus we cannot consider using them in general programs. Thus we limit a quantum computer to be used as Grover Search in our framework. It should also be noted that we accrue many benefits (as we will describe below) from this limitation.

2.3. *Overview of Proposed Framework*

As mentioned above, only Grover Search may be used for general programs among the existing quantum algorithms. However, it is not trivial to use Grover Search in a general program. Thus, we propose an efficient framework to use Grover Search. Our framework is as shown in Fig. 1, and it has the following steps.

- **Step 1.** A programmer writes a program by using standard C++ programming language.
- **Step 2.** He/she inserts a special C++ comment to specify that the evaluation of the if-expression directly after the comment may be done by Grover Search. (This will be explained in detail in Sec. 3.)
- **Step 3.** Our framework automatically generates a corresponding quantum circuit to evaluate the if-expression specified at Step 2. (This step will be explained in detail in Sec. 4.1.)
- **Step 4.** By evaluating the number of primitive quantum gates in the quantum circuit generated at Step. 3, the framework determines whether the

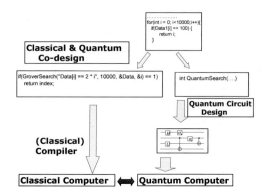

Figure 1.   This figure shows an overview of proposed framework. Some parts of a source program are automatically extracted to be performed on a quantum computer. Corresponding quantum circuits for the parts are automatically generated.

processing time of the quantum circuit is faster than the processing time of the corresponding if-expression on a classical computer. If the framework determines that the quantum circuit is faster, it generates some interface source codes for a classical computer. (This will be explained in Sec. 4.2.)

- **Step 5.** The framework optionally generates a C++ source code that simulates the behavior of the quantum circuit generated at Step. 3. This code can be used for the entire simulation by using only a classical computer as we will explain in Sec. 4.2.

3. Program Part Extraction for Grover Search

Suppose we encounter the part as shown in Fig. 2 in our target program. This part essentially needs 10,000 evaluations of the if-expression on a classical computer. However, if we utilize Grover Search, we need only roughly $\sqrt{10,000}(= 100)$ times of Grover Iterations. However note that a single evaluation of the if-expression on a classical computer is much faster than a single Grover Iteration. Thus, as mentioned in the previous section, our strategy is the following: First, we construct a quantum circuit for the corresponding Grover Iteration by the method described in Section 4.1. Then we estimate the necessary steps, i.e., the number of elementary gates, on a quantum computer to perform Grover Search. Then we can determine whether this part should be performed on a quantum computer.

The above decision can be done automatically without any help from a programmer. However, the following issues cannot be solved without help from a programmer: Grover Search finds *one* solution *at random* among several solutions, and therefore, the behavior of Grover Search is not exactly the same as an if-expression in a for-loop where the first solution in the for-loop always becomes the solution. In other words, Grover Search can be used for cases when a programmer wants to find only one solution, and he/she does not care about the order of the for-loop. This fact is obviously only known to the programmer. Therefore, in our framework a programmer is requested to put a special comment "//Quantum Search" to specify this property before the corresponding if-expression as shown in Fig. 2. This is a minimum load for programmers. Indeed they do not need to know the details of quantum computation, and the entire program can be written in pure C++ unlike other quantum language researches[1,2,3].

4. Compiling the Extracted Parts for Grover Search

For a given C++ program, we can extract a part that should be performed on a quantum computer as mentioned above. Then our framework generates a quantum circuit for a single Grover Iteration corresponding to the part as will be mentioned in Sec. 4.1. It also generates additional C++ source codes to utilize the quantum circuit from a classical computer and deals with Grover Search error, as will be mentioned in Sec. 4.2.

4.1. *Generating a Quantum Circuit for Grover Search*

Here we describe the construction of a quantum circuit fed to a quantum computer. Almost all the proposed quantum circuit design methods are based on decomposing a unitary matrix that corresponds to a target quantum algorithm into elementary gates[7]. Although such circuit design methods can be applied to any quantum algorithm, they cannot be applied to large problems since the size of a unitary matrix is 2^n for quantum algorithms with n qubits. Since program parts for quantum computation should include large size Grover Search (otherwise the parts should be performed on a classical Computer, as mentioned in Sec. 2.1), we cannot utilize unitary matrix decomposition in our framework. Thus we adopt the following design methodology.

- We adopt the structure of the Grover Iteration as shown in Fig. 3. In the figure, W means the Walsh-Hadamard transformation. C_0 is a transformation that inverts the phase of state $|0 \cdots 0\rangle$. Note that their structures are fixed for all problems, and thus we can easily construct them as shown in the figure.
- C_f is a transformation that inverts the phase of state $|x\rangle$ such that x is a binary representation of the solution for Grover Search. This is of-

```
for(int i = 0; i<10000;i++){
//Quantum Search
   if(Data[i] == 2 * i) {
      return i;
   }
}
```

Figure 2.   Example of a for-loop that may
be solved by Grover Search.

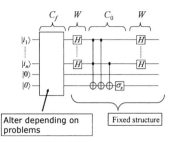

Figure 3.   Quantum circuit for Grover Iteration.

ten called an "oracle" in the literature. This part changes depending on search problems, and thus we need to construct C_f from scratch.

In other words, we do not generate the entire quantum circuit from scratch, but we utilize the known structure of Grover Search as much as possible. Thus, our essential task is to construct a quantum circuit for C_f.

Below we explain how to construct C_f by using an example. Suppose we want to generate a quantum circuit for Grover Search that corresponds to the part, as shown in Fig. 2. Then C_f should work as follows.

- It takes an input quantum state $|i\rangle$ where i is a binary representation of integer i. (This needs $\log 10,000$ qubits.)
- It does not change the quantum states of $|i\rangle$, but transforms one bit ancilla qubit state from $|0\rangle$ to $|1\rangle$ if (Data[i] == 2 * i) is satisfied. Otherwise, it does not change the ancilla qubit.

In our framework, we assume that we have quantum registers (quantum memory) that take $|i\rangle$ as an input and outputs the value of the i-th element in the register into ancilla qubits. This is the most standard model of quantum memory. We can see an example of a quantum register in Fig. 4 where i and $Data[i]$ are encoded by three qubits.

Here we want to implement a function such that it becomes 1 only when (Data[i] == 2 * i). This is a Boolean function with respect to the bits for the binary expressions of i and $Data[i]$. For simplicity, we consider that i and $Data[i]$ are expressed by only three bits. Then our essential task is to construct a quantum circuit for the function with six input variables, and we can construct such a circuit by generalized Toffoli gates as shown in Fig. 4. In the figure, a black circle means the normal control bits whereas a white circle works in the opposite way, i.e., a generalized Toffoli gate works when the qubit state is $|0\rangle$. The function corresponding to (Data[i] == 2 * i) is calculated into the qubit on the bottom. The

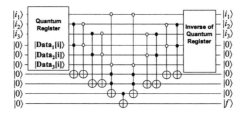

Figure 4. This figure shows an example of how to construct C_f. The left half of the circuit calculates the function into the qubit on the bottom. Then the remaining half of the circuit restores the initial states for the other qubits. i_1 and $Data_1[i]$ are the most significant bits for the encoding.

right half part of the circuit is for resetting ancilla qubits to the initial states. Note that this is necessary when used in Grover Search.

The expression in the example, (Data[i] == 2 * i), is relatively easy, but even for general cases, we can construct C_f as follows:

- **Step 1.** Generate a Boolean function f corresponding to the expression.
- **Step 2.** Generate an exclusive-OR sum of products expression (ESOP) form for f by a standard method in the (classical) logic design[8].
- **Step 3.** Construct a quantum circuit by replacing each product term in the ESOP form at Step. 2 with a generalized Toffoli gate.
- **Step 4.** Convert each generalized Toffoli gate into elementary gates by the method in Ref. 9.

Note that even though the above procedure is not sophisticated, it can be applied to large problems since we only treat the expression and representation of functions, not unitary matrices with exponential size. Therefore, we claim that our framework is practical. We may incorporate optimization into the above procedure.

4.2. Generating Additional C++ Codes

After generating a quantum circuit, our framework automatically modifies the part in the original C++ program. For example, the part as shown in Fig. 2 is converted into the following lines in the original C++ program:

```
if (GroverSearch("Data[i] == 2 * i", 10000, &Data, &i, "i") == 1)
    return i;
```

The function GroverSearch is also defined so that it has the following functionalities: (i) It asks a quantum computer to perform a single Grover Search and receives an answer from the quantum computer. (ii) It verifies the solution. If the solution is incorrect, it continues to ask the quantum computer until a correct answer is obtained or until it has repeated it an appropriate number of times. (This will be mentioned below.) (iii) If a correct solution cannot be obtained within an appropriate number of repetitions, it performs the original C++ source code to find a correct solution. These additional operations are necessary because a single Grover Search has some constant error, i.e., we cannot always get a correct answer by Grover Search. The appropriate number of repetitions are determined as follows. Let the computational cost (on a classical computer) of all the evaluations in a for-loop in the original C++ source code be T_C. Also let the computational cost of a single Grover Search be T_Q. In the following discussion, for simplicity, let us assume the probability that a single Grover Search will succeed to be $3/4$. Then the probability that we cannot find a correct solution after k trials of Grover Search is $(\frac{1}{4})^k$. Thus, the expected total computational cost of the above procedure is $((\frac{1}{4})^k \cdot T_C) + \sum_{l=1}^{k-1}(\frac{3}{4} \cdot (\frac{1}{4})^{l-1} \cdot l \cdot T_Q)$. We can calculate an appropriate value for k to minimize the expected total computational cost, a value which our framework uses in GroverSearch.

Optionally, we can make a C++ source code to simulate the behavior of a quantum circuit, although the simulation on a classical computer may take a long time. If we replace the part that calls a quantum computer for Grover Search with such a code, the entire source code becomes pure C++ language so that we can perform it on a classical computer, i.e., we can simulate the behavior of classical and quantum computers by only using a classical computer.

5. Conclusion

Unlike existing research on quantum programming, our strategy only focuses on Grover Search. By doing so, we have the following advantages: (i) A programmer can write a program by only using (exactly the same) C++ language, and thus the entire program can be simulated by only using current programming environments. (ii) Programmers do not need to worry about which parts should be done by quantum computers. (iii) We can design quantum circuits with a strategy that can be applied to large size problems.

As far as we know, this is the first framework to build a *practical* programming environment for quantum computation. We are now constructing a prototype of our proposed framework and will evaluate it in the future. For other future work, we will consider the possibility of utilizing Amplitude Amplification[10], which is a generalization of Grover Search, in our framework.

References

1. S. Bettelli, T. Calarco, and L. Serafini, *Eur. Phys. J.* **25**, 181 (2003).
2. J. W. Sanders and P. Zuliani, *Math. of Program Construction*, 80 (2000).
3. B. Ömer, *PhD thesis*, Technical University of Vienna (2003).
4. L. K. Grover, *Symp. on Theory of Computing*, 212 (1996).
5. G. De Micheli and M. Sami, *Hardware-Software Co-design*, Kluwer Academic Publishers (1996).
6. Peter W. Shor, *Symp. on Foudations of Computer Science*, 124 (1994).
7. R. R. Tucci, quant-ph/9902062.
8. T. Sasao, *Switching Theory for Logic Synthesis*, Kluwer Academic Publishers (1999).
9. A. Barenco, C. H. Bennett, R. Cleve, D. P. DiVincenzo, N. Margolus, P. Shor, T. Sleator, J. A. Smolin, and H. Weinfurter. *Physical Review A*, **52(5)**, 3457 (1995).
10. G. Brassard, P. Hoyer, M. Mosca, and A. Tapp, quant-ph/0005055.

DECOY STATE QUANTUM KEY DISTRIBUTION

HOI-KWONG LO

Center for Quantum Information and Quantum Control
Department of Physics and Department of Electrical & Computer Engineering
River Edge, NJ 07661, USA
University of Toronto, Toronto, Ontario, Canada
E-mail: hklo@comm.utoronto.ca

Quantum key distribution (QKD) allows two parties to communicate in absolute security based on the fundamental laws of physics. Up till now, it is widely believed that unconditionally secure QKD based on standard Bennett-Brassard (BB84) protocol is limited in both key generation rate and distance because of imperfect devices. Here, we solve these two problems directly by presenting new protocols that are feasible with only current technology. Surprisingly, our new protocols can make fiber-based QKD unconditionally secure at distances over 100km (for some experiments, such as GYS) and increase the key generation rate from $O(\eta^2)$ in prior art to $O(\eta)$ where η is the overall transmittance. Our method is to develop the decoy state idea (first proposed by W.-Y. Hwang in *"Quantum Key Distribution with High Loss: Toward Global Secure Communication"*, Phys. Rev. Lett. 91, 057901 (2003)) and consider simple extensions of the BB84 protocol. This part of work is published in *"Decoy State Quantum Key Distribution"*, http://arxiv.org/abs/quant-ph/0411004.

We present a general theory of the decoy state protocol and propose a decoy method based on only one signal state and two decoy states. We perform optimization on the choice of intensities of the signal state and the two decoy states. Our result shows that a decoy state protocol with only two types of decoy states—-a vacuum and a weak decoy state—asymptotically approaches the theoretical limit of the most general type of decoy state protocols (with an infinite number of decoy states). We also present a one-decoy-state protocol as a special case of Vacuum+Weak decoy method. Moreover, we provide estimations on the effects of statistical fluctuations and suggest that, even for long distance (larger than 100km) QKD, our two-decoy-state protocol can be implemented with only a few hours of experimental data. In conclusion, decoy state quantum key distribution is highly practical. This part of work is published in *"Practical Decoy State for Quantum Key Distribution"*, http://www.arxiv.org/abs/quant-ph/0503005.

We also have done the first experimental demonstration of decoy state quantum key distribution, over 15km of Telecom fibers. This part of work is published in *"Experimental Decoy State Quantum Key Distribution Over 15km"*, http://www.arxiv.org/abs/quant-ph/0503192.

MIXED-STATE ENTANGLEMENT IN THE LIGHT OF PURE-STATE ENTANGLEMENT CONSTRAINED BY SUPERSELECTION RULES

STEPHEN D. BARTLETT

School of Physics,
The University of Sydney, New South Wales 2006, Australia
E-mail: bartlett@physics.usyd.edu.au

HOWARD. M. WISEMAN

Centre for Quantum Computer Technology,
Centre for Quantum Dynamics, School of Science,
Griffith University, Brisbane, 4111 Australia
E-mail: H.Wiseman@griffith.edu.au

ROBERT W. SPEKKENS

Perimeter Institute for Theoretical Physics,
Waterloo, Ontario N2J 2W9, Canada
E-mail: rspekkens@perimeterinstitute.ca

ANDREW C. DOHERTY

School of Physical Sciences,
The University of Queensland, Brisbane, 4072, Australia
E-mail: doherty@physics.uq.edu.au

We show that the classification of bi-partite pure entangled states when local quantum operations are restricted, e.g., constrained by local superselection rules, yields a structure that is analogous in many respects to that of mixed-state entanglement, including such exotic phenomena as bound entanglement and activation. This analogy aids in resolving several conceptual puzzles in the study of entanglement under restricted operations. Specifically, we demonstrate that several types of quantum optical states that possess confusing entanglement properties are analogous to bound entangled states. Also, the classification of pure-state entanglement under restricted operations can be much simpler than for mixed state entanglement. For instance, in the case of local Abelian superselection rules all questions concerning distillability can be resolved.

1. Introduction

Quantum information, if understood as "information about a quantum system", is identical with the quantum state ρ [1]. Thus quantum information science is about the manipulation and measurement of quantum systems, and the corresponding transformations of the information, i.e., of their states. As is well known, the most general transformation of a quantum state has the form of a trace-decreasing completely positive map [2]

$$\rho \rightarrow \mathcal{E}(\rho) \tag{1}$$

where $\mathrm{Tr}[\mathcal{E}(\rho)]$ is the probability that the corresponding manipulation can be performed. If we consider all possible initial states and all possible maps, then everything is determined simply by the dimension of the Hilbert space. This is **boring**!

To make quantum information science interesting, one must place restrictions, typically on the operations. Indeed, Ben Schumacher [3] has put forward a program for doing research in quantum information science, that can be paraphrased as follows:

(1) Consider an interesting *subset* of operations
(2) Explore the resulting information theory
(3) Write a paper
(4) GOTO (1)

The paradigm for such research is to consider the subset of operations that two parties can perform with LOCC. That is, with local operations plus classical communication. Under this restriction, the notion of *entanglement* emerges as a resource that allows for interesting tasks such as disproving local realism [4], performing quantum teleportation [5], or other forms of quantum information processing [2].

For pure states, at least in the bi-partite setting, entanglement is very well characterized. However, in the presence of noise, it is currently not known precisely *which* entangled states are useful, and a vast theory of mixed-state entanglement (MSE) has developed to classify states according to their entanglement properties [6]. In this paper (based upon Ref. [7]) we show that the theory of pure-state entanglement when quantum operations are restricted – described formally by a superselection rule [8,9,10] – precisely replicates the structure of MSE including such exotic properties as bound entanglement and activation. Moreover, unsolved questions in MSE have analogous questions in pure state entanglement under restrictions that can be answered. Thus, there is a deep *analogy* between MSE and pure state entanglement under restrictions that sheds light on both areas.

2. Mixed State Entanglement

Central to the theory of entanglement is the classification of quantum states shared between two parties (Alice and Bob) who can perform only local quantum operations and classical communication (LOCC). One important class is the subset LP of states that are *locally preparable*, that is, preparable by LOCC (starting with some uncorrelated fiducial state). Another is the class of states that are *distillable* [11], denoted D. States in D are such that n copies can be converted into nr pure maximally entangled states via LOCC for some $r > 0$ in the limit $n \to \infty$. A *pure* state is either in LP or in D, depending on whether it is a product state or not (i.e., a state of the form $|\psi\rangle_A \otimes |\phi\rangle_B$ or not). On the other hand, there are mixed states that are neither locally preparable nor distillable: so-called *bound entangled* states [12].

Identifying the class D of mixed states that are distillable is important for quantum information processing, but unfortunately this identification has confounded researchers for many years [6]. A related class, with a simpler characterisation, is the class of states that are *1-distillable* [13,14], denoted 1-D. A state ρ is in 1-D if there exists a quantum operation [2] $\mathcal{E}$ implementable with LOCC that maps ρ onto a 2×2-dimensional space of the bi-partite system such that $\mathcal{E}(\rho)$ is non-separable [a]. In addition, a state ρ is n-distillable (in n-D) if there exists an LOCC operation $\mathcal{E}_n$ onto a 2×2-dimensional space such that $\mathcal{E}_n(\rho^{\otimes n})$ is non-separable. If a state is in n-D for some n then it is in D. In fact, it has been shown [16] that n-D is a proper subset of D for all n. For pure states, 1-distillability is equivalent to distillability. Thus, every pure state is either in LP or in 1-D. Due to the existence of bound entangled states, and the fact that 1-D $\subset$ D, there exist mixed states that are neither in LP nor in 1-D. We shall refer to all such states as *1-bound*.

Remarkably, this proper gap between the classes LP and 1-D consisting of 1-bound states can be removed if we extend the set of operations that Alice and Bob can perform beyond LOCC. We describe this extension of operations as supplementing LOCC with an additional *resource*. Clearly, additional power will affect the boundaries of what Alice and Bob can prepare or distill; we are interested in a resource that precisely removes the gap between LP and 1-D. Consider extending LOCC to allow all operations that preserve the positivity of the partial transpose of states [17]. With this additional resource, all states with positive partial transpose (PPT) can be prepared locally; all states that are not PPT are in 1-D [18]. States that are not locally preparable with LOCC, but locally preparable given LOCC plus the additional resource, can be said to *become* locally preparable given the resource,

---

[a]A *separable* state possesses a convex decomposition into product states. On a 2×2-dimensional space, all separable states are in LP, and all non-separable states are in D [15].

denoted BLP. Similarly, the class of states that are not 1-distillable but become 1-distillable given the resource we denote as B1-D. For mixed bipartite states under PPT-preserving operations, the class BLP contains all PPT bound entangled states and the class B1-D contains all non-PPT states that are not 1-D; both classes are non-empty [12,13,14]. See Fig. 1.

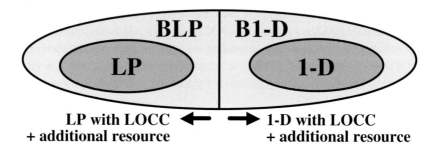

Figure 1.   Illustration of the division of all bipartite mixed states into four classes.

The categories BLP and B1-D are related in an interesting way: there exist states in BLP that can *activate* [19] the entanglement of states in B1-D using only LOCC operations. Any PPT-preserving operation can be implemented probabilistically using LOCC and a specific state in BLP [20]. Thus, for every $\rho \in$ B1-D there exists a state $\sigma \in$ BLP such that $\rho \otimes \sigma$ is in 1-D. Another remarkable feature of MSE is that there exist states in B1-D that are nevertheless distillable [16] (i.e., n-distillable for some $n \geq 2$).

There remain, however, several open questions regarding the general structure of MSE. Perhaps the most important question from the point of view of quantum information processing is: Are all states in B1-D distillable?

3. Pure-State Entanglement under Restrictions

We now show that the above structure of MSE has a precise analogue in the structure of *pure-state* entanglement when the allowed local quantum operations are restricted. The latter structure has the advantage of only describing pure states, which are mathematically much simpler to characterise [b]. We formulate our restriction on operations in the form of a superselection rule (SSR) associated with

---

[b]Note that one could characterise MSE under such restrictions; however, the classification of such states is likely to be at least as difficult as unrestricted MSE.

a finite or compact Lie group G [10,21]. (Alternate methods for the study of entanglement under restrictions can be found in [22].)

The SSR we describe can be defined operationally as follows. Suppose Alice and Bob exchange a pair of systems, described by a Hilbert space $\mathcal{H}^A \otimes \mathcal{H}^B$, with a third party, Charlie. Suppose further that the local reference frames of Alice, Bob and Charlie, which transform via a group G, are uncorrelated: that is, the element $g \in G$ relating Alice's and Charlie's local frames is completely unknown, as is the element $g' \in G$ relating Bob and Charlie's local frames. It follows that a preparation represented by a density matrix ρ on $\mathcal{H}^A$ relative to Alice's frame is represented by the density matrix $\mathcal{G}_A[\rho]$ relative to Charlie's frame, where

$$\mathcal{G}_A[\rho] \equiv \int_G d\nu(g)\, T^A(g)\rho T^{A\dagger}(g), \qquad (2)$$

with $T^A(g)$ a unitary representation of g on $\mathcal{H}^A$, and $d\nu$ the group-invariant (Haar) measure. The operations that Alice can implement relative to Charlie's frame are represented by completely positive maps $\mathcal{O}_A$ that commute with $\mathcal{G}_A$. A similar result holds for the operations that Bob can implement. The joint operations that Alice and Bob can implement relative to Charlie's frame are those represented by maps $\mathcal{O}_{AB}$ that commute with $\mathcal{G}_A \otimes \mathcal{G}_B$. These are said to be locally G-invariant [10]. This restriction on operations is referred to as a local superselection rule for G (local G-SSR).

The local G-SSR then induces the following structure in the local Hilbert spaces (we consider $\mathcal{H}^A$): $\mathcal{H}^A = \bigoplus_q \mathcal{H}_q^A$ splits into "charge sectors" labeled by q and each carrying inequivalent representations $T_q'^A$ of G, and then a further decomposition of each sector into a tensor product

$$\mathcal{H}_q^A = \mathcal{M}_q^A \otimes \mathcal{N}_q^A \qquad (3)$$

of a subsystem $\mathcal{M}_q^A$ carrying an irreducible representation $T_q'^A$ and a subsystem $\mathcal{N}_q^A$ carrying a trivial representation of G (these are the noiseless subsystems relative to $\mathcal{G}_A$ [23]). The action of $\mathcal{G}_A$ on a density operator ρ in terms of this decomposition is

$$\mathcal{G}_A[\rho] = \sum_q \mathcal{D}_{Aq} \otimes I_{Aq}(\Pi_q^A \rho \Pi_q^A), \qquad (4)$$

where Π_q^A is the projection onto the charge sector q, $\mathcal{D}_{Aq}$ is the map that takes every operator on $\mathcal{M}_q^A$ to a constant times the identity operator on that space, and I_{Aq} is the identity map over operators in the space $\mathcal{N}_q^A$. The effect of the local G-SSR, then, is to remove the ability to prepare states or measure operators that have coherence between different local charge sectors or that are not completely mixed over the subsystems $\mathcal{M}_q^A$. The same structure arises for $\mathcal{H}^B$ and provides an analogous decomposition of $\mathcal{G}_B$. For further details, see [10,21].

To classify the entanglement of multiple copies of a state (necessary for addressing questions of distillability), we now demonstrate how to treat multiple systems under a local G-SSR. If the system that Alice exchanges with Charlie is made up of several systems, $\mathcal{H}^A = \otimes_i \mathcal{H}_i^A$, then because all the subsystems are defined relative to Alice's frame, the uncertainty in the element $g \in G$ relating Alice's frame to Charlie's is represented by Eq. (2) using the tensor representation $T^A = \otimes_i T_i^A$.

We now demonstrate that the structure of MSE discussed above has a precise analogue in the structure of pure state entanglement constrained by a G-SSR. The set of LOCC operations that are locally G-invariant will be denoted by G-LOCC. Pure bipartite states are locally preparable under a G-SSR, that is, preparable by G-LOCC, denoted $\text{LP}_{G\text{-SSR}}$, iff (i) the state is a product state, and (ii) it is locally G-invariant. (Thus, not all pure product states are in $\text{LP}_{G\text{-SSR}}$.) A state $|\psi\rangle$ is 1-distillable with G-LOCC, denoted $1\text{-D}_{G\text{-SSR}}$, if there exists an operation $\mathcal{E}$ in G-LOCC mapping $|\psi\rangle$ onto a 2×2-dimensional space such that $\mathcal{E}[|\psi\rangle\langle\psi|]$ is locally G-invariant and non-separable. It follows from the main theorem of Bartlett and Wiseman [10] that $|\psi\rangle$ is in $1\text{-D}_{G\text{-SSR}}$ iff $\mathcal{G}_A \otimes \mathcal{G}_B[|\psi\rangle\langle\psi|]$ is 1-distillable with unrestricted LOCC. Both $\text{LP}_{G\text{-SSR}}$ and $1\text{-D}_{G\text{-SSR}}$ are non-empty; explicit examples of each can be constructed as product/non-product states within 2×2 subspaces or subsystems that are noiseless relative to $\mathcal{G}_A \otimes \mathcal{G}_B$.

As with MSE, there is a proper gap between the classes of pure states that are locally preparable with G-LOCC and those that are 1-distillable with G-LOCC. This class of states, containing both product and non-product pure states, is analogous to the class of 1-bound states in MSE. Moreover, it is possible to extend G-LOCC in such a way that any pure state in this gap becomes either locally preparable or 1-distillable. One simply lifts the restriction of the local G-SSR by providing Alice and Bob with Charlie's local frame, so that the local frames of the three parties are correlated. With this additional resource, Alice and Bob can now implement any operation in LOCC. Extending G-LOCC to LOCC divides the proper gap between $\text{LP}_{G\text{-SSR}}$ and $1\text{-D}_{G\text{-SSR}}$ into two classes, both of which are non-empty. All product states that are not locally G-invariant (i.e., product states not in $\text{LP}_{G\text{-SSR}}$) become locally preparable with G-LOCC given the shared reference frame for G, denoted $\text{BLP}_{G\text{-SSR}}$. This result is obvious because all pure product states are locally preparable with unrestricted LOCC. All non-product states $|\psi\rangle$ for which $\mathcal{G}_A \otimes \mathcal{G}_B[|\psi\rangle\langle\psi|]$ is not 1-D (i.e., non-product states not in $1\text{-D}_{G\text{-SSR}}$) become 1-distillable with G-LOCC given the shared reference frame for G, denoted $\text{B1-D}_{G\text{-SSR}}$. This result is also obvious because all pure non-product states are 1-distillable with unrestricted LOCC. Thus, we have demonstrated that the structure of Fig. 1 for MSE has a precise analogue in the structure of pure state entanglement under the restriction of a G-SSR.

In quantum optics, where a local photon-number U(1)-SSR applies whenever two parties fail to share a phase reference [24], there has been considerable debate over the entanglement properties of several types of states. The debate is resolved by recognizing that operational notions of entanglement that coincided for pure states under unrestricted LOCC, namely being not locally preparable and being 1-distillable, do not coincide under a local U(1)-SSR, and the states in question are judged entangled by one notion and not the other.

Perhaps the most widely-known example is the two-mode single-photon state $|0\rangle_A|1\rangle_B + |1\rangle_A|0\rangle_B$ [25,26]. With unrestricted operations, this state is entangled; however, under a local photon-number U(1)-SSR, the state's entanglement properties are no longer as transparent. The state is not 1-distillable using U(1)-LOCC, and is thus "unentangled" according to one notion of entanglement. However, it is not locally preparable with these restricted operations, and thus "entangled" by a different notion of entanglement. The resolution is that there exist additional categories of pure entangled states under a local U(1)-SSR, analogous to the additional categories arising for MSE. Because the state in question becomes 1-distillable given unrestricted LOCC, it is in B1-$D_{U(1)\text{-SSR}}$. As another example, a class of non-product states of indistinguishable particles have been identified [9] as having "no particle entanglement"; here we would say they have no *1-distillable* entanglement under a local G-SSR.

A related development has been the identification of states which are separable but not locally preparable under a U(1)-SSR [27,8]. Examples of such states are $|+\rangle_A|+\rangle_B$ and $|-\rangle_A|-\rangle_B$, where $|\pm\rangle = |0\rangle \pm |1\rangle$ [c]. Ref. [8] identified such states as a "new type of nonlocal resource"; we classify such states as BLP$_{U(1)\text{-SSR}}$. We see that the remarkable (and often confusing) entanglement properties of states under restricted operations can be understood by recognising that different operational notions of entanglement do not coincide in this case, leaving a structure akin to that of MSE.

4. Particular Results for the U(1)-SSR

As well as illuminating the general structure of MSE, the context of restricted operations for pure states permit simple answers to distillability questions that have proved difficult to answer MSE . As a specific example, we will now present

---

[c]Refs. [27,8] considered states invariant under global U(1) such as the equal mixture of $|+\rangle_A|+\rangle_B$ and $|-\rangle_A|-\rangle_B$. These are simply *mixed* versions of BLP$_{U(1)\text{-SSR}}$ states.

a complete characterization of the activation and distillability properties of any pure state constrained by a local Abelian U(1)-SSR (e.g., a local charge or particle number SSR) [9,24]. Under such a SSR, all operations are locally U(1)-invariant; that is, they commute with $\mathcal{U}_A \otimes \mathcal{U}_B$, where

$$\mathcal{U}_A[\cdot] \equiv \int d\phi \, e^{-i\phi \hat{N}_A}(\cdot) e^{i\phi \hat{N}_A} = \sum_n P_n^A(\cdot) P_n^A, \tag{5}$$

with $\hat{N}_A$ the local number operator and P_n^A the projector onto its nth eigenspace, and where $\mathcal{U}_B$ is defined similarly.

We have shown [7] the following results:

Theorem (activation): For all $|\Psi\rangle \in$ B1-$D_{U(1)\text{-SSR}}$, there exists a $|\chi\rangle \in$ BLP$_{U(1)\text{-SSR}}$ such that $|\Psi\rangle \otimes |\chi\rangle$ is in 1-$D_{U(1)\text{-SSR}}$. We say that $|\chi\rangle$ has *activated* the entanglement in $|\Psi\rangle$.

Theorem (distillation): All pure non-product states are distillable using U(1)-LOCC, with at most three copies of the state required for distillation.

Moreover, there are some states for which three copies are necessary (3-D but not 2-D), some for which two are necessary (2-D but not 1-D), and some which are 1-D. Also we have determine explicit distillation protocols for these two classes of B1-D states (those that are 2-D and those that are not).

5. Conclusion

In summary, we have shown how to reproduce the complex classification of MSE by restricting local operations so as to create a proper gap between what is locally preparable and what is 1-distillable. Debates over the entanglement properties of pure states under restricted operations are resolved by recognizing novel categories of entanglement in this context. Our results suggest that the exotic structure of MSE is generic, and that developing entanglement theory under other sorts of restrictions is a promising direction for further research. They also suggest that it may be fruitful to think of standard LOCC as a restriction relative to the "more natural" PPT-preserving operations, and to consider whether a resource that lifts this restriction might be established with the same ease as a shared reference frame.

References

1. A. Duwell, Stud. Hist. Phil. Mod. Phys. (Sept. 2003) pp. 479-499 (2003).
2. M. A. Nielsen and I. L. Chuang, *Quantum Computation and Quantum Information* (Cambridge University Press, Cambridge, 2000).

3. B. Schumacher, talk presented at the Third Conference of the ESF programme: *Quantum Information Theory and Quantum Computing* (Erice, Sicily, Italy, March 15-22, 2003).
4. J. S. Bell, Physics **1**, 195 (1964).
5. C.H. Bennett, G. Brassard, C. Crepeau, R. Jozsa, A. Peres, and W. K. Wooters, Phys. Rev. Lett. **70**, 1895. (1993)
6. M. Horodecki, P. Horodecki and R. Horodecki, in *Quantum Information: An Introduction to Basic Theoretical Concepts and Experiments*, Vol. 173 of *Springer Tracts in Modern Physics* ed. G. Alber *et al.* (Springer Verlag, Berlin, 2001), pp. 151-195, arXiv:quant-ph/0109124.
7. S. D. Bartlett, A. C. Doherty, R. Spekkens, and H. M. Wiseman, arXiv:quant-ph/0412158.
8. F. Verstraete and J. I. Cirac, Phys. Rev. Lett. **91**, 010404 (2003).
9. H. M. Wiseman and J. A. Vaccaro, Phys. Rev. Lett. , **91**, 097902 (2003).
10. S. D. Bartlett and H. M. Wiseman, Phys. Rev. Lett. **91**, 097903 (2003).
11. C. H. Bennett, G. Brassard, S. Popescu, B. Schumacher, J. A. Smolin and W. K. Wootters, Phys. Rev. Lett. **76**, 722 (1996); C. H. Bennett, D. P. DiVincenzo, J. A. Smolin and W. K. Wootters, Phys. Rev. A **54**, 3824 (1996).
12. M. Horodecki, P. Horodecki and R. Horodecki, Phys. Rev. Lett. **80**, 5239 (1998).
13. D. P. DiVincenzo, P. W. Shor, J. A. Smolin, B. M. Terhal and A. V. Thapliyal, Phys. Rev. A **61**, 062312 (2000).
14. W. Dür, J. I. Cirac, M. Lewenstein and D. Bruß, Phys. Rev. A **61**, 062313 (2000).
15. M. Horodecki, P. Horodecki and R. Horodecki, Phys. Rev. Lett. **78**, 574 (1997)
16. J. Watrous, Phys. Rev. Lett. **93**, 010502 (2004).
17. E. M. Rains, IEEE Trans. Inf. Theory **47**, 2921 (2001).
18. T. Eggeling, K. G. H. Vollbrecht, R. F. Werner, and M. M. Wolf, Phys. Rev. Lett. **87**, 257902 (2001).
19. P. Horodecki, M. Horodecki and R. Horodecki, Phys. Rev. Lett. **82**, 1056 (1999).
20. J. I. Cirac, W. Dür, B. Kraus and M. Lewenstein, Phys. Rev. Lett. **86**, 544 (2001).
21. A. Kitaev, D. Mayers and J. Preskill, Phys. Rev. A **69**, 052326 (2004).
22. H. Barnum, E. Knill, G. Ortiz, and L. Viola, Phys. Rev. A **68**, 032308 (2003); H. Barnum, E. Knill, G. Ortiz, R. Somma and L. Viola, Phys. Rev. Lett. **92**, 107902 (2004).
23. E. Knill, R. Laflamme and L. Viola, Phys. Rev. Lett. **84**, 2525 (2000).
24. B. C. Sanders *et al.*, Phys. Rev. A **68**, 042329 (2003).
25. S. M. Tan, D. F. Walls, and M. J. Collett, Phys. Rev. Lett. **66**, 252 (1991).
26. L. Hardy, Phys. Rev. Lett. **73**, 2279 (1994).
27. T. Rudolph and B. C. Sanders, Phys. Rev. Lett. **87**, 077903 (2001).

QUANTUM COMPUTATION BASED ON ELECTRON SPIN QUBITS WITHOUT SPIN-SPIN INTERACTION

YIN-ZHONG WU[†‡*], WEI-MIN ZHANG[†], AND CHOPIN SOO[†]

[†] *Department of Physics, and Center for Quantum Information Science, National Cheng Kung University, Tainan 70101, Taiwan.*
[‡] *Department of Physics, Changshu Institute of Technology, Changshu 215500, P. R. China.*

Using electron spin states in a unit cell of three semiconductor quantum dots as qubit states, a scalable quantum computation scheme is advocated without invoking qubit-qubit interactions. Single electron tunneling technology and coherent quantum-dot cellular automata architecture are used to generate an ancillary charge entangled state which is then converted into spin entangled state. Without using charge measurement and ancillary qubits, we demonstrate universal quantum computation based on free electron spin and coherent quantum-dot cellular automata.

1. Introduction

Schemes utilising spin degrees of freedom in quantum computation are favorably supported in recent experiments showing unusually long spin decoherence time in semiconductors[1,2]. It has been known that universal quantum computations based on electron spins can be achieved by tunable Heisenberg type spin-spin interactions[3,4], or XY interactions[5,6]. A few years ago, Knill et al. [7] showed that a universal quantum computation can be realized for photons using only linear optics and single-photon detectors without qubit-qubit interactions. Pauli's exclusion principle and the nature of fermion operators imply the situation for free fermions is quite different from that of bosons[8,9]. Recently, Beenakker et al. [10] showed that for free-flying fermions, one is able to construct a controlled-NOT gate, using only beam splitters and single spin rotations if a single charge measurement is added. However, an interaction-free mechanism for logical operations on electron spins is still lacking in solid state systems. Our proposal uses electrostatic gates to control the electron's position; and an ancillary charge entangled state is generated with the help of a coherent quantum-dot cellular automata structure. The charge entangled state is then converted into spin entangled state, and

---

*E-mail: yzwu@cslg.cn

a controlled NOT operation is obtained by using only single spin rotations. Spin-spin interactions are neither required nor invoked in our scheme. Our proposal is a demonstration of free electron spin quantum computation in semiconductor nanostructure devices.

2. Basic unit

The basic device in our architecture is based on a semiconductor quantum-dot array as shown in Fig. 1. Three quantum dots (e.g. 1-A-B, 2-C-D) are constructed as a unit cell, and gate electrodes are integrated in each quantum dot. The solid lines in each cell indicate the possibility of interdot tunneling. The confining barrier between two cells must ensure that tunneling of electrons between different cells is ideally forbidden, or at least highly suppressed. There exists only one excess conductor electron in each cell; and spin states of the excess electron are chosen as qubit states. Note that four empty quantum dots of neighboring unit cells (e.g.

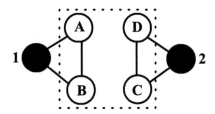

Figure 1.   Schematic drawing of our CNOT gate composed of two unit cells(each cell contains three QDs). There is only one excess conductor electron in each cell. The lines in each unit cell indicate the possibility of interdot tunneling. The four empty quantum dots of neighboring unit cells (e.g. ABCD) form a usual coherent quantum-dot cellular automata structure(dotted square box).

ABCD) form a coherent quantum-dot cellular automata(CQCA) structure(the dotted square box shown in Fig. 1). Quantum-dot cellular automata(QCA) introduced here has been used to simulate the classical digital algorithms[12]. When the QCA cell is charged with two excess electrons, two electrons will occupy diagonal sites as a result of Coulomb repulsion, we define the two diagonal polarization states as $|+\rangle$ and $|-\rangle$ for the two full polarizations $P = 1$ and $P = -1$ respectively. This is shown in Fig. 2(a). A semiconductor realization of such a structure has actually been developed[13]. The structure of CQCA was proposed by G. Toth[14]; and the charge state can be in an arbitrary superposition state of the two diagonal polarizations $\alpha|+\rangle + \beta|-\rangle$, which is different from that in QCA where only the full polarization states can be reached. In order to do a coherent operation, two input parameters(γ and P_{bias}) within a CQCA unit are designed[14]. The tunneling energy

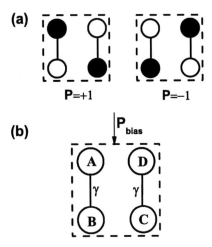

Figure 2. (a)Two full polarization charge states of the two electrons within the quantum-dot cellular automata caused by Coulomb repulsion. (b) Coherent operations on the two full polarization states system by an external bias polarization. γ is the tunneling energy between vertical QDs, and P_{bias} denotes the bias polarization.

γ is set by external electrodes that lower or raise the interdot barrier between the vertical quantum dots, and the bias polarization P_{bias} can be realized by bias gate voltages. The effective interaction between the external bias polarization and the charge state of two electrons within CQCA can be expressed as $E_0 P_{bias} \sigma_z$[14], where E_0 is the strength of Coulombic coupling of two electrons, and the eigenstates of σ_z are the two full polarization states. We will show later that only a relative phase between the two full polarization states is required in our scheme. So it is not necessary to tune the tunneling energy between vertical QDs in our CNOT gate. Turning on P_{bias}, a relative phase between the two full polarization states will be reached if $E_0 P_{bias} \gg \gamma$. Thus, only one input parameter P_{bias} in CQCA is needed to implement our CNOT gate(See Fig. 2(b)).

3. Definition of qubit states

The electron state is a direct product of charge and spin states $|e_i\rangle|S_i\rangle(i = 1, 2)$. Only electron spin states are chosen as qubit states. Initially the excess conductor electrons of the two neighbor unit cells reside in quantum dot 1 and 2. The excess electron in each cell is driven away from its initial position only when a two-qubit operation is performed, and the electrons will return their positions as soon as the two-qubit operation is completed. To wit the electron position in a unit cell initially is the charge"ground state" $|e_{g_i}\rangle$, where subscript i stands for the particular unit cell. We can take the charge state $|e_i^0\rangle(|e_i^1\rangle)$ as the state of ith electron sitting at the upper(lower) position in the cellular automata. Three steps are needed to realize a controlled NOT operation on the spin state of the two electrons. In the first step, an ancillary charge entangled state is generated. Then, four single spin rotations are performed on individual QDs A, B, C, and D. Finally, by pushing the two electrons to their initial positions, we realize a controlled NOT operation on the spin states. Each step will be described in more detail in the following section.

4. Implementation of universal quantum computation

Firstly, turning on the gate voltage to lower the site energy on dots A, B, C, and D, the electron in site 1 tunnel into quantum dots A and B. If two quantum dots A and B are identical, the probabilities of electron in site 1 tunneling into quantum dots A and B will be equal. At the same time, the electron in site 2 will similarly tunnel into quantum dots C and D. According to the property of cellular automata structure(based on Coulomb repulsion principle), the two electrons will tend to reside on the two diagonal positions in the four quantum dots(A, B, C, and D) with equal probability. After this process, we obtain a charge Bell state of the two electrons $\frac{1}{\sqrt{2}}(|e_1^0e_2^1\rangle + |e_1^1e_2^0\rangle)$. Then we turn on the bias polarization by bias gate voltages. If the duration of electric pulses satisfy $\int E_0 P_{bias}(t)dt = \frac{3\pi}{4}$, we can realized a phase factor on the above charge Bell state, and an ancillary charge entangled state, i.e.,$\frac{1}{\sqrt{2}}(i|e_1^0e_2^1\rangle + |e_1^1e_2^0\rangle)$, is obtained. Quantum computation schemes using two full polarization states in quantum-dot cellular automata as qubit states have been proposed before[14]. In contrast to these works, here we choose electron spin states as qubit states, and the ancillary charge entangled state is used only as an intermediate state. Up to now, we have done nothing on the electron spin states(qubit states).

We shall next demonstrate the implementation of a controlled NOT operation based on the ancillary charge entangled state and only on rotations of single spins. To wit, four single spin operations $R_z(\frac{\pi}{2})$, $R_z(\frac{3\pi}{2})$, $R_x(\frac{\pi}{2})$,and $R_x(\frac{3\pi}{2})$ are performed on individual quantum dots A, B, C, and D, respectively. Spin rotations $R_z(\theta)$ and $R_x(\theta)$ are defined as $e^{-i\theta Z/2}$ and $e^{-i\theta X/2}$, respectively. After this second step, the state of two electrons becomes $\frac{1}{\sqrt{2}}[iR_z(\frac{\pi}{2})|S_1\rangle R_x(\frac{\pi}{2})|S_2\rangle|e_1^0e_2^1\rangle + R_z(\frac{3\pi}{2})|S_1\rangle R_x(\frac{3\pi}{2})|S_2\rangle|e_1^1e_2^0\rangle]$.

The third step is basically the reverse process of the initial procedure. Turning on the gate voltage bias reversely, electron 1 and electron 2 will return their initial positions in real space, and the charge entangled state collapses into its ground state $|e_1^0e_2^1\rangle \longrightarrow |e_{g_1}\rangle|e_{g_2}\rangle$, $|e_1^1e_2^0\rangle \longrightarrow |e_{g_1}\rangle|e_{g_2}\rangle$. The ancillary charge entangled state is converted into a spin entangled state. Now the spin and charge states of the two electrons become $\frac{1}{\sqrt{2}}[iR_z(\frac{\pi}{2})|S_1\rangle R_x(\frac{\pi}{2})|S_2\rangle + R_z(\frac{3\pi}{2})|S_1\rangle R_x(\frac{3\pi}{2})|S_2\rangle]|e_{g_1}\rangle|e_{g_2}\rangle = \frac{1}{\sqrt{2}}[iR_z(\frac{\pi}{2}) \otimes R_x(\frac{\pi}{2}) + R_z(\frac{3\pi}{2}) \otimes R_x(\frac{3\pi}{2})]|S_1\rangle|S_2\rangle|e_{g_1}\rangle|e_{g_2}\rangle = [U_{CNOT}|S_1\rangle|S_2\rangle]|e_{g_1}\rangle|e_{g_2}\rangle$, wherein the CNOT operation acts on the spin states(qubit states), and an irrelevant overall phase factor has been ignored. We can further implement a universal quantum gate by combining the Controlled-NOT gate with single spin rotations. The end result of these operations is that universal quantum computation can be achieved.

In our scheme, the single qubit rotation is easily realized via a local magnetic field or all-optical control of the electron spin with near-field technology. R_x and R_z single spin rotations can be implemented by tuning on the local magnetic field along x direction and z direction respectively. However, achieving one-qubit operations by applying magnetic fields is rather slow. To decrease the operation time, two laser pulses with near-field technology can be employed to generate single spin rotations. To wit, pulsed lasers with different polarizations are addressed on a quantum dot. We use Δ_1 and Δ_2 to denote the detuning of the two lasers. $\Omega_1(\Omega_2)$ denotes the Rabi frequency between the spin-up state $|0\rangle$(the spin-down state $|1\rangle$) and the excited state of the quantum dot. Ω_1 and Ω_2 can be enhanced by increasing the intensity of the laser field. For large detuning and under Raman resonance($\Delta = \Delta_1 = \Delta_2 \gg \Omega_{eff}$), i.e., adiabatically eliminating the excited state[15], an effective interaction between the spin-up state $|0\rangle$ and spin-down state $|1\rangle$ can be realized: $\Omega_{eff}(t)(|0\rangle\langle 1|e^{-i[(\omega_1-\omega_2)t+\delta\phi_{12}]} + |1\rangle\langle 0|e^{i[(\omega_1-\omega_2)t+\delta\phi_{12}]})$, where $\Omega_{eff}(t) = \frac{\Omega_1(t)\Omega_2(t)}{\Delta}$, and $\delta\phi_{12}$ is the initial phase difference of the two lasers. By properly adjusting the duration and the phase of each laser pulse to satisfy $\int_0^T \Omega_{eff}(t)dt = \frac{\theta}{2}$ and $(\omega_1 - \omega_2)T + \delta\phi_{12} = 2n\pi$, a single spin rotation $R_x(\theta)$ is achieved. The single spin rotation $R_y(\theta)$ can be obtained under the condition of $\int_0^T \Omega_{eff}(t)dt = \frac{\theta}{2}$ and $(\omega_1 - \omega_2)T + \delta\phi_{12} = 2n\pi + \frac{3\pi}{2}$. Combining single spin rotation R_x with R_y, we can obtain an arbitrary rotation along z direction $R_z(\theta)$, for instance, single spin rotation $R_z(\frac{\pi}{2})$ on quantum dot A can be achieved by $R_y(\frac{3\pi}{2})R_x(\frac{3\pi}{2})R_y(\frac{\pi}{2})$, and single spin rotation $R_z(\frac{3\pi}{2})$ on quantum dot B can be replaced by $R_y(\frac{3\pi}{2})R_x(\frac{\pi}{2})R_y(\frac{\pi}{2})$. Recently, ultrafast manipulation of electron spins on femtosecond time scales has been attained by Gupta et al[17] (optical "tipping" pulses can enact substantial rotation of electron spin through a mechanism utilizing the optical Stark effect). In our construction, due to the identical, or very similar, energy spectra of QDs A and B(also C and D), it is needed, or at least desirable, to employ near-field technology for laser pulses to manipulate single spin operations, but this constraint can be removed if only two single spin operations are designed within the CQCA structure[18].

5. Further remarks

A static uniform magnetic field can be employed to split the spin-up state and the spin-down state by Zeeman energy for the initialization. The measurement of the spin state can be realized by converting spin information to charge information, which in turn can be made to depend on the spin state. For example, Martin et al.[19] proposed a scheme for single spin readout using an electron trap near a conducting channel. Recently, Friesen et al.[20] propose a novel spin-charge transduction scheme which converts spin information to orbital information within a single QD by microwave excitation, and their scheme can be used for both rapid initialization and readout on electron spin states in a semiconductor quantum dot.

The structure shown in Fig. 1 can also be implemented by two polygonal(triangular for the case here) quantum dots. Further control of the electron location may be achieved by placing electrodes at the vertices and applying the requisite voltages to generate the desired potential well for the electron. The control of electrons in a polygonal quantum dot had been examined in detail elsewhere[21]. If such a quantum dot can be manufactured with only one excess electron, the operation time of controlling electron location in the space-time will decrease significantly. Recently, single electron tunneling through triple virtual quantum dots has been studied experimentally[22]. Although such a scheme based on solid QD arrays would be technologically challenging, it yields a new way to implement electron spin entangled state without resorting to qubit-qubit interactions. Besides, such structures can be fabricated in a scalable fashion. The separation between each CNOT gate structure should be much greater than the size of the cell to ensure that the Coulomb repulsion between electrons on different CNOT gate structures is always much smaller than that between the two electrons within the same cellular automata structure.

6. Conclusion

In summary, a charge entangled state is generated in space-time based on a semiconductor coherent quantum-dot cellular automata. This is then converted into a spin entangled state. The controlled-NOT operation can be implemented in fermion systems without resorting to spin-spin interactions. Combining with single qubit operations, free electron spin quantum computation can be realized based on solid nanostructures. Our proposal does not require charge detection and the ancillary qubit, and is easily scalable. All the operations are performed only on individual quantum dots. The major technological challenges in this scheme are sufficiently fast control in SET and the precision of near-field technology in pulsed lasers.

Acknowledgments

One of the authors (Y. Z. Wu) would like thank Prof. Yan-Ten Lu, and Prof. Chuan-Pu Liu for useful discussions. This work has been supported by the National Center for Theoretical Sciences, Taiwan, and the Natural Science Foundation of Educational Committee, Jiangsu Province; and by the National Science Council of ROC under grant nos. NSC-93-2120-M-006-005, NSC-93-2112-M-006-011, NSC-93-2112-M-006-019, NSC93-2112-M-006.

References

1. D. D. Awschalom, D. Loss, and N. Samarth, *Semiconductor Spintronics and Quantum Computation*, (Springer-Verlag Berlin Heidelberg, 2002).
2. Rogerio de Sousa, and S. Das Sarma, *Phys. Rev. B* **67**, 033301(2003); B. Beschoten, E. Johnston-halperin, D. K. Young, M. Poggio, J. E. Grimaldi, S. Keller, S. P. DenBaars, U. K. Mishra, E. L. Hu, and D. D. Awschalom, *Phys. Rev. B* **63**, 121202(R)(2001).
3. D. Loss, and D. P. DiVincenzo, *Phys. Rev. A* **57**, 120 (1998).
4. D. P. DiVincenzo, D. Bacon, J. Kempe, G. Burkard, and K. B. Whaley,*Nature* **408**, 339 (2000).
5. D. A. Lidar, and L. A. Wu, *Phys. Rev. Lett.* **88**, 017905 (2002).
6. A. Imamoglu, D. D. Awschalom, G. Burkard, D. P. DiVincenzo, D. Loss, M. Sherwin, and A. Small, *Phys. Rev. Lett.* **83**, 4204(1999).
7. E. Knill, R. Laflamme, and G. J. Milburn, *Nature* **409**, 46 (2001).
8. B. M. Terhal, and D. P. DiVincenzo, *Phys. Rev. A* **65**, 032325 (2002).
9. E. Knill, quant-ph/0108033.
10. C. W. J. Beenakker, D. P. DiVincenzo, C. Emary, and M. Kindermann, *Phys. Rev. Lett.* **93**, 020501(2004).
11. A. O. Orlov, I. Amlani, G. H. Bernstein, C. S. Lent, and G. L. Snider, *Science* **277**, 928(1997).
12. Islamshah Amlani, Alexei O. Orlov, Geza Toth, Gary H. Bernstein, Craig S. Lent, and Gregory L. Snider, *Science* **284**, 289(1999).
13. Y. Fu, and M. Willander, *J. Appl. Phys.* **83**, 3186(1998); M. Govemale, M. Macucci, G. lannaccone, and C. Ungarelli, *J. Appl. Phys.* **85**, 2962(1999); M. Macucci, M. Gattobigio, and G. Iannaccone, *J. Appl. Phys.* **90**, 6428(2001);
14. Geza Toth, and Craig S. Lent, *Phys. Rev. A* **63**, 052315(2001).
15. Nikolay V Vitanov, Thomas Halfmann, Bruce W Shore, and Klaas Bergmann, *Annu. Rev. Phys. Chem.* **52**, 763(2001).
16. H. Grabert, and M. H. Devoret, *Single charge tunneling:coulomb blockade phenomena in nanostructures*, (Plenum Press, New York, 1992).
17. J. A. Gupta, R. Knobel, N. Samarth, D. D. Awschalom, *Science* **292**, 2458 (2001).
18. See also, for instance, W. M. Zhang, Y. Z. Wu and Chopin Soo, *Free spin quantum computation with semiconductor nanostructures*, quant-ph/0502002.
19. I. martin, D. Mozyrsky, and H. W. Jiang, *Phys. Rev. Lett.***90**, 018301(2003).
20. Mark Friesen, Charles Tahan, Robert Joynt, and M. A. Eriksson, *Phys. Rev. Lett.***92**,

037901(2004).

21. C. E. Creffield, Wolfgang Hausler, J. H. Jefferson, and Sarben Sarkar, *Phys. Rev. B***59**, 10719(1999); J. H. Jefferson, M. Fearn, and D. J. Tipton, *Phys. Rev. A***66**, 042328(2002).

22. A. Vidan, R. M. Westervelt, M. Stopa, M. Hanson, and A. C. Gossard, *Appl. Phys. Lett* **85**, 3602(2004).

INFRARED WAVELENGTH QUANTUM COMMUNICATIONS BASED ON SINGLE ELECTRON TRANSISTORS

DAVID M. T. KUO[†]

Department of Electrical Engineering,
National Central University, ChungLi, Taiwan,
320 Republic of China

We propose to employ a selective formation method to embed an isolated self assembled quantum dot into a n-i-n junction to implement single electron transistors (SETs). The absorption and emission spectrum of SETs are theoretically studied by the Keldysh Green function method. The electronic levels and Coulomb interactions of electrons of InAs quantum dot (QD) are evaluated by an effective mass model. It is found that Coulomb interaction and level mixing in the many body open system lead to double peaks associated with the intraband transitions involving two lowest levels of the QD. We can electrically control the SETs as a single-photon source and double-photon source for $10 \mu m$ wavelength. The single photon source can be used in the application of quantum cryptograph, and the double-photon source can be utilized in the teleportation.

KEYWORDS: quantum dots, single electron transistors, Keldysh Green's function.

The quantum dot (QD) system has many potential applications in electronic devices, including quantum dot lasers[1] and infrared detectors[2]. Recently, the spontaneous emission spectrum of QDs embedded in a semiconductor p-n junction has been proposed as a generator of single photon, which is important in the application of quantum cryptograph[3]. The anti-bunching feature of a single photon source was demonstrated by optical and electrical pumping, respectively[4-5]. So far, the implementation of a single photon source has focused on short wavelengths (near $1 \mu m$). For such short wavelength photon, it could be generated by the electron-hole recombination in the excitations including exciton, negative trion, positive trion and biexciton states[6].

In some applications of quantum communications, long wavelength single photon at near $10 \mu m$ infrared wavelength is useful due to its advantage of high

---

[†] Electronic mail: mtkuo@ee.ncu.edu.tw

transmission on atmosphere. To generate a long wavelength single photon, we can employ electron intraband transitions. The idea to realize such photon source is to use single electron transistors (SETs), which is consisted of two leads and one QD. The SETs not only can be used as a single photon detector, but also a single photon emitter. The mechanism of these applications is based on the Coulomb blockade effect of SETs with only few bound states in a dot.

Up to now, many nano-structure materials such as the silicon (Si) and germanium (Ge) QDs can be made very small by advanced fabrication technologies. Futhermore, most recently the Ge/Si SETs have been demonstrated at room temperature[7,8]. Nevertheless, we propose to embed a single self-assembled InAs QD into GaAs slab (with width W) which is connected with two dimensional quantum well electrodes, because Si and Ge are indirect band gap semiconductors. The studied system is shown by Fig. 1. In the fabrication of optoelectronic devices such as single photon generators and single photon detectors, the adopted materials prefer the direct band gap semiconductors. Even though it is difficult to align a single dot with the source and drain electrodes, this difficulty may be solved by the selective formation method[9,10], which can improve position control and minimize the size fluctuation of self-assembly quantum dot.

To produce a single photon at near $10\mu m$ infrared wavelength, first we need to know the energy levels of InAs QDs, which depends on the shape of QDs. Here the cylindrical InAs QDs with the radius R_0 and height $h = 2R_0$ will be considered. Within the effective mass model, the Hamiltonian is

$$(-\nabla \frac{\hbar^2}{2m_e^*(\rho,z)}\nabla + V_{QD}^e(\rho,z) - eFz)\psi(\rho,z,\phi) = E\psi(\rho,z,\phi), \qquad (1)$$

where $m_e^*(\rho,z)$ denotes the position-dependent electron effective mass, which takes $m_G^* = 0.067m_e$ for GaAs and $m_I^* = 0.04m_e$ for InAs. Due to the combined effect of strain and nonparabolicity, the electron effective mass of InAs in QDs is larger than that of bulk ($0.024m_e$). $V_{QD}^e(\rho,z)$ is approximated by a constant potential $V_0 = 0.5eV$ in the QDs region (this includes the effect of hydrostatic strain due to the lattice mismatch between IsAs and GaAs). eFz term arises from the applied bias, where F denotes the strength of the electric field. Besides the energy levels of InAs QDs, the Coulomb interactions of electrons in the InAs QDs are unavoidable for small QDs. We expect that the Coulomb interactions will influence spectrum significantly. Particle interaction (by taking direct Coulomb interaction into account only) is calculated by

$$U_{i,j} = \frac{e^2}{\varepsilon_0} \int dr_1 dr_2 \frac{n_i(r_1) n_j(r_2)}{|r_1 - r_2|} \qquad (2)$$

where $n_i(r_1) = |\psi_i(\rho, z, \phi)|^2$ is the particle density of ith energy level. For the purpose of constructing the approximate wave functions, we place the system in a large cylindrical confining box with length L and radius R (L and R must be much larger than the length and radius of the cylindrical InAs QDs). Here we adopt $L = 60nm$, $R = 40nm$. The wave functions are expanded in a set of basis functions, which are chosen to be products of Bessel functions and sine waves

$$\psi_{n,l,m}(\rho, z, \phi) = J_l(\beta_n \rho) e^{il\phi} \sin(k_m(z + L/2))$$

where $k_m = m\pi/L, m = 1,2,3...$ J_l is the Bessel function of order of l and $\beta_n R$ is the nth zero of J_l. We solve the eigenfunctions of the effective-mass Hamiltonian via the Ritz variational method. Figure 1 shows the lowest three energy levels of the confined states as a function of the QD size. Figure 2 shows the intralevel Coulomb interactions (U_{11} and U_{22}) and interlevel Coulomb interaction U_{12} as a function of the QD size. The strengths of Coulomb interactions are, in general, inversely proportional to the QD size, since the charge densities in smaller QDs are more localized. However as the QD size decreases below a threshold value (around $R_0 = 3.6nm$), the Coulomb interactions U_{22} and U_{12} become reduced as a result of the leak out of charge density of the first excited state.

After the calculation of the energy levels and Coulomb interactions, we construct a model that can be used to simulate the bias dependent light emission and absorption processes. For the nonequilibrium system considered here, it is convenient to use the Keldysh's Green function method to calculate the transport and optical properties[11,12].We solve the Anderson Hamiltonian of a two-level system coupled with leads in the presence of an electromagnetic radiation with frequency ω. The tunneling current is given by

$$J = \frac{2e}{h} \sum_j \int d\varepsilon [f_L(\varepsilon - \mu_L) - f_R(\varepsilon - \mu_R)] \frac{\Gamma_{j,L}\Gamma_{j,R}}{\Gamma_{j,L} + \Gamma_{j,R}} \mathrm{Im}\, G_j^r(\varepsilon) \qquad (3)$$

where $f_L(\varepsilon)$ and $f_R(\varepsilon)$ are the Fermi distribution function for the left lead and right lead, respectively. The chemical potential difference between these two leads is related to the applied bias $\mu_L - \mu_R = eV_a$. $\Gamma_{j,L}$ and $\Gamma_{j,R}$ denote the tunneling rates from the QD to the left lead and right lead, respectively, for electrons in level j. Notations e and h denote the electron charge and Planck's constant. The calculation of tunneling current is entirely determined by the spectral function $A_j(\varepsilon) = \mathrm{Im}\, G_j^r(\varepsilon)$, which is the imaginary part of the retarded Green function G_j^r. The expression of G_j^r can be obtained by the equation of motion method. We treat particle interaction with Hartree-Fock approximation, which is appropriate for the case of Coulomb blockade[10]. After a lengthy algebra, the retarded Green's function is given by

$$
G_i^r(\varepsilon) = \frac{(1-N_i)(1-N_j)^2}{\varepsilon - E_i + i\dfrac{\Gamma_i}{2}} + \frac{2(1-N_i)(1-N_j)N_j}{\varepsilon - E_i - U_{12} + i\dfrac{\Gamma_i}{2}} + \frac{(1-N_i)N_j^2}{\varepsilon - E_i - 2U_{12} + i\dfrac{\Gamma_i}{2}} ,
$$

$$
+ \frac{N_i(1-N_j)^2}{\varepsilon - E_i - U_{ii} + i\dfrac{\Gamma_i}{2}} + \frac{2N_iN_j(1-N_j)}{\varepsilon - E_i - U_{ii} - U_{12} + i\dfrac{\Gamma_i}{2}} + \frac{N_iN_j^2}{\varepsilon - E_i - U_{ii} - 2U_{12} + i\dfrac{\Gamma_i}{2}}
$$

(4)

Note that $i \neq j$ in Eq. (4). N_i is electron occupation number of ith energy level. Γ_i is the sum of Γ_{iR} and Γ_{iL}. The retarded Green's function of Eq. (4) $G_i^r(\varepsilon)$ contains an admixture of six possible configurations in which a given electron can propagate. The number of the electrons in the QD can be solved in a self-consistent way

$$
N_i = \frac{1}{\pi} \int d\varepsilon \frac{\Gamma_{i,L} f_L(\varepsilon) + \Gamma_{i,R} f_R(\varepsilon)}{\Gamma_{i,L} + \Gamma_{i,R}} \mathrm{Im}\, G_i^r(\varepsilon)
$$

$$
= \frac{1}{\pi} \int d\varepsilon\ f_i^<(\varepsilon) \mathrm{Im}\, G_i^r(\varepsilon)
$$

(5)

The value of N_i is restricted in the region $0 \leq N_i \leq 1$. In Eqs. (3), (4) and (5) we ignored the effect of weak electron-photon interaction. This means that the photon assisted tunneling currents were neglected in Eq. (3).

Although the expression of Eq. (4) is complex, it can be reduced to a simple form for the narrow band case, which is defined as $E_F \ll \Delta E = E_2 - E_1$, where E_F is the Fermi energy of lead. In the narrow band case, the energy levels E_1 and E_2 will not be covered simultaneously by either μ_L or μ_R or μ_L and μ_R. Therefore, the tunneling current is determined by the $G_j^r(\varepsilon)$ with only two poles $\varepsilon = E_j$ and $\varepsilon = E_j + U_{jj}$. Their weighting factors are $(1 - N_j)$ and N_j, respectively. To generate a single photon at $10\,\mu m$ wavelength, we are seeking an intraband transition at an energy around

124 meV. Because long wavelength infrared single photon passes through the atmosphere in the 3-5 μm band and 8-14 μm band[13], furthermore the atmospheric transmission is near 98% in the infrared range from 9.7 μm to 10.3 μm. The QD with radius from $R_0 = 5.4nm$ to $R_0 = 5.8nm$ can be adopted to generate a single photon at near 10 μm wavelength.

From the results of Fig. 1, the QD with radius $R_0 = 5.6nm$ is sufficient to provide such photons and gives the following physical parameters: $E_1 = 143meV$, $E_2 = 265meV$, $U_{11} = 23meV$, $U_{12} = 20.8meV$ and $U_{22} = 19.8meV$. To simulate the system, we also choose $E_F = 30meV$ which is 10 meV below E_1 , $\Gamma_1 = 0.01meV$ and $\Gamma_2 = 0.3meV$, respectively. Note that the tunneling rates are assumed bias independent for simplicity, even though Γ can be determined with a reliable method[14]. The tunneling current (not shown here) displays a staircase behavior. Two plateaus are generated by the intralevel Coulomb interactions U_{11} and U_{22} , which correspond to the charging energy of the ground state and the first excited state. This is the well-known Coulomb blockade effect[12].

When the chemical potential $\mu_{L/R}$ sweeps through the ground state, the detected photon spectrum is described by the imaginary part of polarization $P(\omega)$, which is calculated with non-diagonal parts of the lesser Green function $G_{12}^<(\varepsilon)$,

$$\text{Im} P(\omega) = \lambda^2 \int \frac{d\varepsilon}{\pi} f_1^<(\varepsilon)(1 - f_2^<(\varepsilon + \omega))\{\text{Im}(\frac{1 - N_1}{\varepsilon - E_1 + i\frac{\Gamma_1}{2}}) \text{Im}(\frac{-1}{\varepsilon - E_2 + \omega - i\frac{\Gamma_2}{2}})$$

$$+ \text{Im}(\frac{N_1}{\varepsilon - E_1 - U_{11} + i\frac{\Gamma_1}{2}}) \text{Im}(\frac{-1}{\varepsilon - E_2 - U_{12} + \omega - i\frac{\Gamma_2}{2}})\} \quad (6)$$

where λ is the Rabi frequency. Figure 3 shows the absorption spectrum as a function of frequency for various applied voltages at zero temperature: solid line ($V_a = 24mV$) and dashed line ($V_a = 70mV$). At low voltage ($E_F + eV_a/2 < E_1 + U_{11}$), electrons in the QD can only occupy the single-particle state; thus, the spectrum displays a single peak at resonance frequency, $\omega = E_2 - E_1$. As voltage further increases to satisfy $E_F + eV_a/2 > E_1 + U_{11}$, the spectrum displays two peaks centered at frequency $\omega = \Delta E_{21} = E_2 - E_1 = 122meV$ and

$\omega = E_2 - E_1 + U_{12} - U_{11} = 119.8 meV$. The high frequency peak corresponds to the interlevel transition [with relative probability $(1 - N_1)$] of a single electron in the QD, while the low frequency peak corresponds to the interlevel transition [with a relative probability N_1] of a second electron in the QD under the influence of the first electron. Thus, the effect of electron correlation leads to a double-peak spectrum with energy separation related to the intralevel and interlevel Coulomb energies ($\Delta U_{21} = U_{12} - U_{11} = -2.2 meV$).

When the lead supplies electrons into the first excited state, light emission process occurs. The emission spectrum is given by

$$\text{Im} P(\omega) = \lambda^2 \int \frac{d\varepsilon}{\pi} f_2^<(\varepsilon)(1 - f_1^<(\varepsilon - \omega))\{\text{Im}(\frac{1 - N_2}{\varepsilon - E_2 + i\frac{\Gamma_2}{2}}) \text{Im}(\frac{1}{\varepsilon - E_1 - \omega - i\frac{\Gamma_1}{2}})$$

$$+ \text{Im}(\frac{N_2}{\varepsilon - E_2 - U_{22} + i\frac{\Gamma_2}{2}}) \text{Im}(\frac{1}{\varepsilon - E_1 - U_{12} - \omega - i\frac{\Gamma_1}{2}})\}$$

(7)

Figure 4 shows the emission spectrum for the different applied voltages: solid line ($V_a = 268 mV$) and dashed line ($V_a = 310 mV$). The solid line exhibits a single peak at $\omega = E_2 - E_1 = 122 meV$, because the applied bias is not sufficient to overcome the Coulomb bloackade. When the applied bias overcomes the Coulomb blockade and supplies the second electron into the first excited state, the spectrum exhibits two peaks centered at frequency $\omega = E_2 - E_1 = 122 meV$ and $\omega = E_2 - E_1 + U_{22} - U_{12} = 121 meV$. Because applied bias will establish electric field, it is worthy studying how the Stark effect to influence the emission spectrum. The Stark effect can be ignored for the absorption spectrum since the applied bias is small. When the applied bias is $V_a = 310 mV$ and the *GaAs* slab width is *30nm*, the electric field strength is $F \approx 1. \times 10^5 V / cm$. We found that the Stark effect can not be neglected in such field strength. The dotted line includes Stark shift for energy levels. The Stark effect leads to a blueshift ($\approx 4 meV$) in the transition energies. Note that the Coulomb interactions are changed very slightly for $F \approx 1. \times 10^5 V / cm$.

In this study it is found that the light transition between the ground state and the first excited state could generate double peaks due to the effect of electron correlation in open system. We can use the tunneling current as functions of bias and double peaks shown in Figs. (3) and (4) to determine the intralevel Coulomb interactions (U_{11} and U_{22}) and the interlevel Coulomb interaction U_{12} . In addition, we can electrically control the SETs as a single photon source and two photons source for infrared wavelength. The single photon source can be used in the application of quantum cryptograph, and the two photons source can be utilized in the teleportation[15].

Acknowledgments

This work was supported by National Science Council of Republic of China Contract Nos. NSC-93-2215-E-008-014 and NSC- 93-2120-M-008-002

References

1. V. I. Klimov, A. A. Mikhailovsky, S. Xu, A. Malko, J. A. Hollingworth, C. A. Leatherdale, H. J. Eisler, and M. G. Bawend, Science **290**, 314 (2000).
2. D. M. T. Kuo, G. Y. Guo and Y. C. Chang, Appl. Phys. Lett. **79**, 3851 (2001).
3. A. Imamoglu and Y. Yamamoto, Phys. Rev. Lett. **72**, 210 (1994).
4. P. Michler, A. Kiraz, C. Becher, W. V. Schoenfeld, P. M. Petroff, L. D. Zhang, E. Hu and A. Imamoglu, Science **290**, 2282 (2000).
5. Z. L. Yuan, B. E. Kardynal, R. M. Stevenson, A. J. Shield, C. J. Lobo, K. Cooper, N. S. Beattlie, D. A. Ritchie, and M. Pepper, Science **295**, 102 (2002).
6. D. M. T. Kuo and Y. C. Chang, Phys. Rev. B **69**, 041306 (2004).
7. P. W. Li, W. M. Lia, D. M. T. Kuo, S. W. Lin, P. S. Chen, S. C. Lu and M. J. Tsai, Appl. Phys. Lett. **85**, 1532 (2004).
8. M. Saitoh, N. Takahashi, H. Ishikuro and T. Hiramoto, Jpn. J. Appl. Phys. **40** 2010 (2001).
9. C. K. Hahn, Y. J. Park. E. K. Kim, S. K. Min, S. K. Jung and J. H. Park, Appl. Phys. Lett.**73**, 2479 (1998).
10. C. K. Hahn, J. Motohisa and T. Fukui, Appl. Phys. Lett. **76**, 3947 (2000).
11. A. P. Jauho, N. S. Wingreen, and Y. Meir, Phys. Rev. B **50**, 5528 (1994).
12. H. Haug and A. P. Jauho, *Quantum Kinetics in Transport and Optics of Semiconductor* (Springer, Heidelberg, 1996).
13. E. Tow and D. Pan, IEEE J. Sel. Top. Quantum Electron. 6, 408 (2000)
14. D. M. T. Kuo and Y. C. Chang, Phys. Rev. B **61**, 11051 (1994).
15. C. Macchiavello, G. M. Palma, and A. Zeilingger, *Quantum Computation and Quantum information theory.* (World. Scientific, Singapore, 1999)

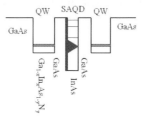

Figure 1. The schematic band diagram for single electron transistor consisted of one InAs QD and two dimensional quantum well electrodes.

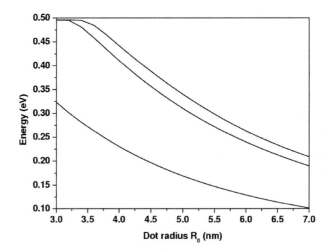

Figure 2. Energies of the bound states of a cylindrical InAs/GaAs QD as a function of the QD size.

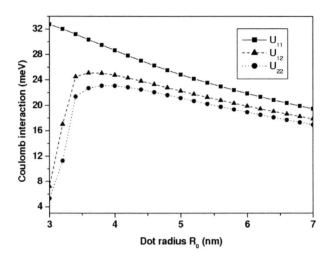

Figure 3. U_{11}, U_{12} and U_{22} as a function of the QD size.

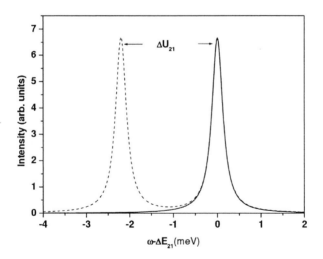

Figure 4. Absorption spectrum as a function of frequency for different applied voltages:solid line ($V_a = 24mV$) and dashed line ($V_a = 70mV$). $\Delta U_{21} = U_{12} - U_{11}$

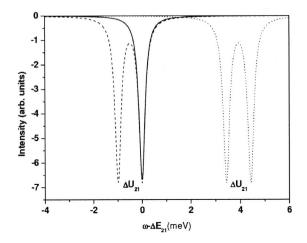

Figure 5. Emission spectrum as a function of frequency for different applied voltages:solid line ($V_a = 268mV$) and dashed line ($V_a = 310mV$). $\Delta U_{21} = U_{22} - U_{12}$.

RECENT RESULTS
IN EXPERIMENTS WITH JOSEPHSON QUBITS

OLEG ASTAFIEV, YURI PASHKIN, TSUYOSHI YAMAMOTO,
YASUNOBU NAKAMURA, JAWSHEN TSAI

Institute for Physical and Chemical Research,
Wako-shi, Saitama, 351-0198, Japan

NEC Fundamental Research Laboratory, 34 Miyukigaoka,
Tsukuba, Ibaraki 305-8501, Japan

We demonstrate our recent experiments on Josephson charge qubits: coherent manipulation of two electrostatically coupled qubits, controlled-NOT operation, single-shot readout, and study of decoherence mechanisms with help of the single-shot readout.

1. Introduction

The Josephson qubit is one of the strongest candidates for building blocks of quantum computers because of their potential scalability. After the first demonstration of coherent manipulation with two charge states in solid-state devices done by Nakamura et. al. in 1999,[1] a considerable progress has been achieved. Coherent manipulation of charge and flux states in different qubit geometries has been realized.[2-5] Recently, an important step towards integration of the qubits has been made: coherent control of two qubits[6] and conditional gate operation[7] have been experimentally demonstrated for two electrostatically coupled charge qubits. To develop multi-qubit integrated circuits the qubits must satisfy a number of strict requirements: among them decoupling from a readout system during the coherent state manipulation and weak intrinsic decoherence caused by electric environment. Understanding of qubit decoherence mechanisms becomes of great importance. In particular, it has been recently shown that the charge fluctuators in a substrate produce dephasing as well as energy relaxation of the Josephson charge qubits.[8, 9]

2. Coherent manipulation with two-qubits

One of the physical realisations of a solid-state qubit is provided by a Cooper-pair box.[10] The two charge states of the box, say $|0\rangle$ and $|1\rangle$, differing by one Cooper pair are coherently mixed by the Josephson coupling as was confirmed experimentally.[11, 12] Quantum state manipulation of such a system can be done by using a non-adiabatic pulse technique, and the read-out can be performed by a properly biased probe electrode. Here we make one step further on the way to implementation of quantum logic gates by integrating two charge qubits and demonstrating their interaction.

The two charge qubits of our circuit are electrostatically coupled by an on-chip capacitor (Fig. 1). The right qubit has a SQUID geometry to allow the control of the Josephson coupling to its reservoir. Both qubits have a common pulse gate but separate dc gates, probes and reservoirs. The pulse gate has nominally equal coupling to each box. The Hamiltonian of the system in the two-qubit charge basis $|00\rangle$, $|01\rangle$, $|10\rangle$ and $|11\rangle$ reads:

$$
H = \begin{bmatrix}
E_{00} & -\dfrac{1}{2}E_{J1} & -\dfrac{1}{2}E_{J2} & 0 \\[2mm]
-\dfrac{1}{2}E_{J1} & E_{10} & 0 & -\dfrac{1}{2}E_{J2} \\[2mm]
-\dfrac{1}{2}E_{J2} & 0 & E_{01} & -\dfrac{1}{2}E_{J1} \\[2mm]
0 & -\dfrac{1}{2}E_{J2} & -\dfrac{1}{2}E_{J1} & E_{11}
\end{bmatrix}
\tag{1}
$$

where $E_{n1n2} = E_{c1}(n_{g1}-n_1)^2 + E_{c2}(n_{g2}-n_2)^2 + E_m(n_{g1}-n_1)(n_{g2}-n_2)$ is the total electrostatic energy of the system (n_1, $n_2 = 0$, 1 is the number of excess Cooper pairs in the first and the second box), E_{J1} (E_{J2}) is the Josephson coupling energy of the first (second) box and the reservoir, $E_{c1,2} = 4e^2 C_{\Sigma2,1}/2(C_{\Sigma1}C_{\Sigma2} - C_m^2)$ are the effective Cooper-pair charging energies, $C_{\Sigma1,2}$ are the sum of all capacitances connected to the corresponding island including the coupling capacitance C_m, $n_{g1,2} = (C_{g1,2}V_{g1,2} + C_pV_p)/2e$ are the normalised charges induced on the corresponding qubit by the dc and pulse gate electrodes. The coupling energy E_m depends not only on C_m, but also on the total capacitance of the boxes: $E_m = 4e^2 C_m/(C_{\Sigma1}C_{\Sigma2} - C_m^2)$. Application of gate voltages allows us to control diagonal elements of the Hamiltonian (1). The circuit was fabricated to have the following relation between the characteristic energies: $E_{J1,2} \sim E_m < E_{c1,2}$. This ensures coherent superposition of the four charge states $|00\rangle$, $|01\rangle$, $|10\rangle$ and $|11\rangle$ around $n_{g1} = n_{g2} = 0.5$ while other charge states are separated by large energy gaps. The above condition justifies the use of a four-level approximation for the description of the system. In our notation $|n_1n_2\rangle$ of the charge states used throughout the text, n_1 and n_2 refer to the number of excess Cooper pairs in the first and the second qubits, respectively.

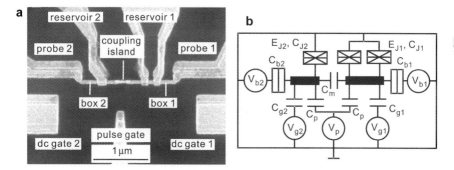

Figure 1. Two capacitively coupled charge qubits. (a), Scanning electron micrograph of the sample. The qubits were fabricated by electron-beam lithography and three-angle evaporation of Al (light areas) on a SiN_x insulating layer (dark). Two qubits are coupled by an additional coupling island overlapping both Cooper-pair boxes. Although the coupling island has a finite tunnelling resistance ~ 10 MΩ to the boxes, we consider the coupling as purely capacitive represented by a single capacitor in the equivalent circuit because all the tunnelling processes are completely blocked. The estimated capacitance of the island to the ground is ~ 1 fF. (b), Equivalent circuit of the device. The parameters obtained from the dc measurements are: $C_{J1} = 620$ aF, $C_{J2} = 460$ aF, $C_{b1} = 41$ aF, $C_{b2} = 50$ aF, $C_{g1} = 0.60$ aF, $C_{g2} = 0.61$ aF, $C_p ≈ 1$ aF, $C_m = 34$ aF, and the corresponding energies are $E_{c1} = 484$ μeV (117 GHz in frequency units), $E_{c2} = 628$ μeV (152 GHz) and $E_m = 65$ μeV (15.7 GHz). Josephson coupling energies, $E_{J1} = 55$ μeV (13.4 GHz) and $E_{J2} = 38$ μeV (9.1 GHz), were determined from the single qubit measurements described later in the text. Probe junction tunnel resistance is equal to 31.6 MΩ (left) and 34.5 MΩ (right). Superconducting energy gap is 210 μeV. Black bars denote Cooper-pair boxes.

In the absence of Josephson coupling, the ground-state charging diagram (n_1, n_2)[13] (see Fig. 2(a)) consists of hexagonal cells whose boundaries delimit two neighboring charge states with degenerate electrostatic energies. For example, points R and L correspond to a degeneracy between the states $|00\rangle$ and $|10\rangle$ and the states $|00\rangle$ and $|01\rangle$ differing by one Cooper pair in the first and the second Cooper-pair box, respectively. If we choose the dc gate charges n_{g1} and n_{g2} far from the boundaries but within the (0,0) cell, then because of large electrostatic energies we can assume that the system remains in the state $|00\rangle$. Since the pulse gate has equal coupling to each qubit, the application of a pulse shifts the state of the system on this diagram along the line tilted at 45 degrees indicated by arrows in Fig. 2(a). The charging diagram remains valid for the small Josephson coupling except on the boundaries where charge states become superposed. When the system is driven non-adiabatically to the point R or L, it behaves like a single qubit oscillating between the degenerate states with a frequency $\omega_{1,2} = E_{J1,2}/\hbar$. Applying arrays of pulses and measuring oscillations of the probe currents I_1 and I_2, we can determine the Josephson energies of each qubit. The accuracy of the measured $E_{J1,2}$ is very high since the electrostatic coupling through C_m has almost no effect on $\omega_{1,2}$ along the boundaries in the vicinity of R and L.

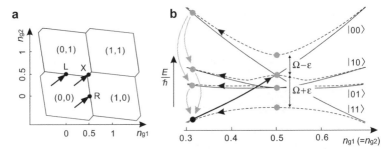

Figure 2. Pulse operation of the device. (a), Schematics of the ground-state charging diagram of the coupled qubits as a function of the normalised gate charges n_{g1} and n_{g2}. The number of Cooper pairs n_1 and n_2 in the neighbouring cells differs by one. The electrostatic energies E_{n1n2} are degenerate at the boundaries. Points R and L correspond to energy degeneracy in the first and the second qubit, respectively. Point X is doubly degenerate: $E_{00} = E_{11}$ and $E_{10} = E_{01}$. Arrows show how pulses shift the system in the experiment. (b), Energy diagram of the system along the line $n_{g1} = n_{g2}$. Solid lines are the electrostatic energies of charge states $|00\rangle$, $|10\rangle$, $|01\rangle$ and $|11\rangle$. Dashed lines are eigenenergies of the Hamiltonian (1). Far from co-resonance (point X in a), the system stays in $|00\rangle$. After the pulse brings the system to the co-resonance (solid arrow), the system starts to evolve producing a superposed state $|\psi(t)\rangle = c_1|00\rangle + c_2|10\rangle + c_3|01\rangle + c_4|11\rangle$. The amplitudes $|c_i|$ (i = 1, 2, 3, 4) remain "frozen" after the pulse termination (dashed arrows) until the resulting state decays into the ground state. The decay process indicated by grey arrows contributes to the probe currents proportional to the probabilities (3).

At the "co-resonance" point X ($n_{g1} = n_{g2} = 0.5$), the system has double degeneracy, $E_{00} = E_{11}$, $E_{10} = E_{01}$, and the dynamics of the quantum evolution becomes more complex and reflects the coupling between the qubits. The cross-section of the energy bands through the point X is shown in Fig. 2(b). Exactly at the co-resonance, all four charge states are mixed and the state of the system can be expressed in general as

$$|\psi(t)\rangle = c_1|00\rangle + c_2|10\rangle + c_3|01\rangle + c_4|11\rangle, \qquad (2)$$

where $|c_i|$ (i = 1, 2, 3, 4) are the time dependent probability amplitudes obeying a normalization condition. Using the Hamiltonian (1) and initial conditions one can calculate the probabilities $|c_i|^2$ of each charge state. However, in our read-out scheme, we measure a probe current $I_{1,2}$ proportional to the probability $p_{1,2}(1)$ for each qubit to have a Cooper pair on it regardless of the state of the other qubit, i.e., $I_1 \propto p_1(1) \equiv |c_2|^2 + |c_4|^2$ and $I_2 \propto p_2(1) \equiv |c_3|^2 + |c_4|^2$. Assuming the initial state at $t = 0$ is $|00\rangle$, we can derive for an ideal rectangular pulse shape of a length Δt the time evolution of these probabilities:

$$p_{1,2}(1) = (1/4)[2 - (1 - \chi_{1,2})\cos\{(\Omega + \varepsilon)\Delta t\} - (1 + \chi_{1,2})\cos\{(\Omega - \varepsilon)\Delta t\}] \quad (3)$$

where $\chi_{1,2} = (E^2_{J2,1} - E^2_{J1,2} + E_m^2/4)/(4\hbar^2\Omega\varepsilon)$, $\Omega = ((E_{J1} + E_{J2})^2 + (E_m/2)^2)^{1/2}/2\hbar$, $\varepsilon = ((E_{J1,2} - E_{J2,1})^2 + (E_m/2)^2)^{1/2}/2\hbar$.

One can see that unlike a single qubit case, there are two frequencies present in the oscillation spectrum of the qubits: $\Omega + \varepsilon$ and $\Omega - \varepsilon$, both dependent on E_{J1}, E_{J2} and E_m. We can identify these frequencies with the energy gaps in Fig. 2(b). Note that in the uncoupled situation ($E_m = 0$), $\Omega \pm \varepsilon = E_{J1,2}/\hbar$ and each qubit oscillates with its own frequency $\omega_{1,2}$. Let us stress, however, that the above consideration is valid only in the ideal case when the pulse has zero rise/fall time and the time evolution occurs exactly at the co-resonance point.

The idea of the experiment is shown schematically in Fig. 2(b). From the state $|00\rangle$ (shown as a black dot), the pulse (solid arrow) brings the system to the co-resonance, and the system evolves for the pulse duration time Δt, producing a superposed state (2) indicated by grey circles. After the pulse termination (dashed arrows), the system remains in the superposed state until it decays (dotted arrows) in the ground state by emitting quasiparticles into the probe junctions biased at $V_{b1,2} \approx 600$ μeV. To accumulate a signal, a pulse array ($\sim 3\times10^5$ pulses) was applied to the pulse gate. The repetition time between the pulses was 64 ns, long enough (in comparison to the quasiparticle relaxation time ~ 10 ns) to let the system decay through a Josephson-quasiparticle cycle[14] and give rise to a probe current proportional to $p_{1,2}$. The estimated amplitude of the applied pulses is $V_p \approx 30$ mV.

The results obtained in this way are presented in Fig. 3. First, by tuning n_{g1} and n_{g2}, we do single qubit measurements by bringing the system to the point R or L and thus exciting autonomous oscillations in one of the qubits (Fig. 3(a)). The oscillations can be fitted to a cosine function with an exponential decay time of about 2.5 ns. The oscillations spectra (right panels of Fig. 3(a)) obtained by the Fourier transform contain one pronounced component at 13.4 GHz for the first qubit and at 9.1 GHz for the second one. We identify these values with E_{J1} and E_{J2}. Judging from our previous experiments we can conclude that these values are close to what we expect for the given fabrication parameters, i.e., overlap area and oxidation conditions. Then, by changing n_{g1} and n_{g2}, the system is driven to the co-resonance and the induced quantum oscillations are traced using the same technique. The oscillation pattern becomes more complex (Fig. 3(b)) and more frequency components appear in the spectrum. The observed spectral properties of the oscillations agree with the predictions of Eq. (3) in a sense that there are two peaks in the spectrum and the peak positions are close to the expected frequencies $\Omega + \varepsilon$ and $\Omega - \varepsilon$ for the parameters $E_{J1} = 13.4$ GHz and $E_{J2} = 9.1$ GHz measured in the single qubit experiments (Fig. 3(a)), and $E_m = 15.7$ GHz estimated from the independent dc current-voltage characteristics measurements. The expected from equation (3) position of the $\Omega + \varepsilon$ and $\Omega - \varepsilon$ peaks is indicated by arrows and dotted lines. The decay time ~ 0.6 ns of the coupled oscillations is shorter compared to the case of independent oscillations as should be expected since an extra decoherence channel appears for each qubit after coupling it to its neighbour. The amplitudes of the spectral peaks, however, do not exactly agree with equation (3). We attribute this to the non-ideal pulse shape (finite rise/fall time ~ 35 ps), and the fact that a small shift of n_{g1} and n_{g2} off the co-resonance drastically changes oscillation pattern. Also, even far from the co-resonance, we still have a small contribution to the initial state from the other than $|00\rangle$ charge states distorting the oscillations. We have performed numerical simulation of the oscillation pattern taking into account a realistic pulse shape and not pure $|00\rangle$ initial condition assuming the system is exactly at the co-resonance. The resulting fits are shown in Fig. 3(b) as solid lines. We found that $E_m = 14.5$ GHz, close to the value estimated from the dc measurements, gives better agreement of the fit with the experimental data.

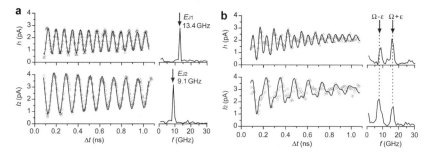

Figure 3. Quantum oscillations in qubits. (a), Probe current oscillations in the first (top) and the second (bottom) qubit when the system is driven to the points R and L, respectively. Right panel shows corresponding spectra obtained by the Fourier transform. In both cases, the experimental data (open triangles and open dots) can be fitted to a cosine dependence (solid lines) with an exponential decay with 2.5 ns time constant. (b), Probe current oscillations in the qubits at the co-resonance point X. Their spectra (right panel) contain two components. Arrows and dotted lines indicate the position of $\Omega + \varepsilon$, $\Omega - \varepsilon$ obtained from (3) using $E_{J1}=13.4$ GHz, $E_{J2}=9.1$ GHz measured in the single qubit experiments (Fig. 3(a)) and $E_m=15.7$ GHz estimated independently from dc measurements. Solid lines are fits obtained from numerical simulation with the parameters $E_{J1}=13.4$ GHz, $E_{J2}=9.1$ GHz and $E_m=14.5$ GHz. Finite pulse rise/fall time and not pure $|00\rangle$ initial condition were taken into account. The introduced exponential decay time is 0.6 ns.

Finally, we checked the dependence of the oscillation frequencies on E_{J1} controlled by a weak magnetic field (up to 20 Gs). The results are shown in Fig. 4. The plot contains the data from both qubits represented by open triangles (first qubit) and open circles (second qubit). Without coupling ($E_m = 0$), the single peaks in each qubit would follow dashed lines with an intersection at $E_{J1} = E_{J2}$. The introduced coupling modifies this dependence by creating a gap and shifting the frequencies to higher and lower values, the spacing between the two branches being equal to $E_m/2h$ when $E_{J1} = E_{J2}$. We compare the observed dependence with the prediction of equation (3) given by solid lines and find a remarkable agreement.

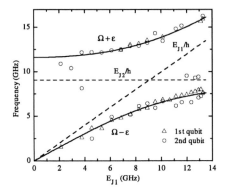

Figure 4. E_{J1} dependence of the spectrum components obtained by the Fourier transform of the oscillations at the co-resonance. Open triangles: frequency components measured in the first qubit; open circles: frequency components measured in the second qubit; solid lines: dependence of $\Omega + \varepsilon$ and $\Omega - \varepsilon$ obtained from (3) using E_{J2}=9.1 GHz and E_m=14.5 GHz and varying E_{J1} from zero up to its maximum value of 13.4 GHz; dashed lines: dependence of the oscillation frequencies of both qubits in the case of zero coupling ($E_m = 0$).

The observed quantum coherent dynamics of coupled qubits in the vicinity of the co-resonance (in particular, double-frequency structure of the probability oscillations in both qubits and frequency "repulsion" at $E_{J1} \approx E_{J2}$ - see Figs. 3(b) and 4) indicates the two qubits become entangled during the course of coupled oscillations although direct measurement of the degree of entanglement was not possible. Simple calculation based on the standard expression for the entanglement of the pure states[15] show that with an ideal pulse shape and the $|00\rangle$ initial condition, the wavefunction (2) passes in its evolution through maximally entangled state in the case of equal Josephson energies. The amount of entanglement does not decrease significantly when realistic experimental conditions are taken into account that is confirmed by the numerical simulations. The relatively large observed oscillation amplitude (about 50% of the expected value) also suggests the existence of entangled states even in our multi-pulse averaged experiment.

3. Controlled-NOT operation

The two-qubit system is the simplest device, which allows to demonstrate quantum logic gats. Here we demonstrate conditional gate operation controlled-NOT based on the two-qubit system. The device shown on Fig. 5(a) is essentially same as that one shown on Fig. 1 with one important modification: two pulse gates are made to separately address each qubit.

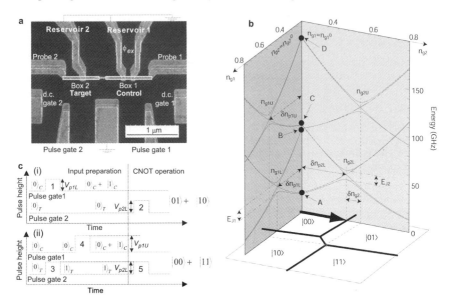

Figure 5. Pulse operation of the coupled-qubit device. (a), Scanning electron micrograph of the sample. The qubits were fabricated by electron-beam lithography and three-angle evaporation of Al on a SiN_x insulating layer above a gold ground plane on the oxidized Si substrate. The two strips enclosed by red lines are the Cooper-pair boxes, which are coupled by an on-chip capacitor.[16] ϕ_{ex} represents magnetic flux penetrating the SQUID loop. An electrode between two pulse gates is connected to the ground to reduce the cross capacitance. Although there is a finite cross capacitance between one gate and the other box (about 15% of the main coupling), it does not play any essential role in the present experiment and so we can neglect it in this paper. The sample was cooled to 40 mK in a dilution refrigerator. The characteristic energies of this sample estimated from the d.c. current-voltage measurements are $E_{c1} = 580$ μeV, $E_{c2} = 671$ μeV and $E_m = 95$ μeV. From the pulse measurements, E_{J1} is found to be 45 μeV at a maximum and E_{J2} to be 41 μeV. The superconducting energy gap is 209 μeV. Probe junction tunnel resistance is equal to 48 MΩ (left) and 33 MΩ (right). (b), Energy band diagram along two lines of $n_{g1} = n_{g1}^{0}$ and $n_{g2} = n_{g2}^{0}$, where n_{g1}^{0} and n_{g2}^{0} are constants. Here $(n_{g1}^{0}, n_{g2}^{0}) = (0.24, 0.26)$, corresponding to the actual experimental condition. In the energy band diagram, lines show four eigenenergies and four electrostatic energies of the states shown in the cells of the charging diagram of the base plane with the corresponding colour. (c), Pulse sequences used in the experiment. In both sequences, the upper and lower patterns show the pulse patterns applied to Pulse gates 1 and 2, respectively. The expected quantum states after each pulse are also shown. The symbols $|0\rangle$ or $|1\rangle$ with subscripts C and T mean the state of the control and target qubits, respectively.

Figure 5(b) represents the idea for the gate operation. Using the Hamiltonian of Eq. (1), we calculate the eigenenergies of the two-qubit system and plot them in the planes $n_{g1} = n_{g1}^{0}$ and $n_{g2} = n_{g2}^{0}$, where n_{g1}^{0} and n_{g2}^{0} are constants. In these planes, if (n_{g1}^{0}, n_{g2}^{0}) is sufficiently far away from the co-resonant point[16] (0.5, 0.5), four energy bands can be regarded as two pairs of nearly independent single-qubit energy bands. In the plane of $n_{g1} = n_{g1}{}^0$, for example, our system is divided into a pair of independent two-level systems $|00\rangle$, $|01\rangle$ and $|10\rangle$, $|11\rangle$. Importantly, the charging energies of each of the two-level systems degenerate at different n_{g2}, namely, at n_{g2L} for the states $|00\rangle$ and $|01\rangle$ and at n_{g2U} for the states $|10\rangle$ and $|11\rangle$ as shown in Fig. 5(b). This difference (δn_{g2}) originates from the electrostatic coupling between the qubits and is given as $E_m/2E_{c2}$. Similarly, we define n_{g1L} and n_{g1U} as shown in the plane of $n_{g2} = n_{g2}{}^0$.

Now we consider the pulse operation. Applying pulses to Pulse gate 1 (2) shifts the system non-adiabatically in the plane of $n_{g2} = n_{g2}{}^0$ ($n_{g1} = n_{g1}{}^0$). For convenience, we define the distances from $(n_{g1}{}^0, n_{g2}{}^0)$ to the degeneracy points as follows: $\delta n_{p1L} = n_{g1L} - n_{g1}{}^0$, $\delta n_{p1U} = n_{g1U} - n_{g1}{}^0$ and $\delta n_{p2L} = n_{g2L} - n_{g2}{}^0$. Suppose we start from the $|00\rangle$ state (point **A**) and apply an ideal rectangular pulse with an amplitude $V_{p2L} = 2e\,\delta n_{p2L}/C_{p2}$ to Pulse gate 2, where C_{p2} is the capacitance between Pulse gate 2 and Box 2. This pulse is represented by the arrow in the ground-state charging diagram of the base plane. In this case, the system is brought to the degeneracy point n_{g2L} and evolves during a pulse duration Δt with a frequency $\Omega = E_{J2}/\hbar$ between the $|00\rangle$ and the $|01\rangle$ states: $\cos(\Omega\Delta t/2)\,|00\rangle +$ $\sin(\Omega\Delta t/2)\,|01\rangle$. By adjusting Δt so that $\Omega\,\Delta t = \pi$ (π pulse), we can stop the evolution when the system is in the $|01\rangle$ state. The system is finally in the state at point **C** after the termination of the pulse.

On the other hand, if we start from the $|10\rangle$ state (point **B**) and apply the same pulse, the system does not reach the degeneracy point for states $|10\rangle$ and $|11\rangle$ (n_{g2U}). In this case, the amplitude of the oscillation between the $|10\rangle$ and the $|11\rangle$ states is suppressed by $E_{J2}{}^2/(E_m{}^2 + E_{J2}{}^2)$. If E_m is sufficiently large, the state $|10\rangle$ remains almost unchanged (except for the phase factor), coming back to point **B** after the termination of the pulse. Similarly, we can realize the transition from the $|01\rangle$ state to the $|00\rangle$ state by the same pulse, and suppress the transition out of the $|11\rangle$ state. Therefore, conditional gate operation can be carried out based on this operation pulse: the target bit is flipped only when the control bit is $|0\rangle$.

To experimentally demonstrate the above gate operation, we prepare different input states from the ground state $|00\rangle$ by applying pulses and measure the output of the gate operation. Figure 5(c) shows two pulse sequences that are utilized in the present experiment. For convenience, each of the pulses in the sequences is labelled by an index m ($m = 1, \ldots, 5$), which we will refer to as "Pulse m". In sequence (i) of Fig. 5(c), a superposition of the states $|00\rangle$ and $|10\rangle$ is created by applying Pulse 1 with the amplitude $V_{p1L} = 2e\,\delta n_{p1L}/C_{p1}$, where C_{p1} is the capacitance between Pulse gate 1 and Box 1. In sequence (ii) of Fig. 5(c), a superposition of the states $|01\rangle$ and $|11\rangle$ is created by two sequential pulses. First, Pulse 3, the same pulse as that for the gate operation, brings the system to the $|01\rangle$ state at point C. Then, Pulse 4 with amplitude $V_{p1U} = 2e\delta n_{p1U}/C_{p1}$ is applied.

In both sequences, an operation pulse (Pulse 2 or 5) creating an entangled state $(\alpha|01\rangle + \beta|10\rangle$ or $\alpha|00\rangle + \beta|11\rangle)$ is applied after the preparation pulses. To change the coefficients α and β, we change the Josephson energy of the control qubit E_{J1} by a magnetic field, while keeping the pulse lengths constant. Because the control qubit has SQUID geometry, E_{J1} is periodically modulated as $E_{J1} = E_{J1max}|\cos(\pi \phi_{ex}/\phi_0)|$, where E_{J1max} is the maximum value of E_{J1} and ϕ_0 is the flux quantum. By repeatedly applying the sequential pulses (with a repetition time $T_r = 128$ ns), we measure the pulse-induced currents through Probes 1 and 2, which are biased at ~ 650 μV to enable a Josephson-quasiparticle (JQP) cycle.[14] These currents are proportional to the probability of the respective qubit having one extra Cooper pair.[16, 17]

We measure ϕ_{ex} dependence of I_C and I_T for pulse sequence (ii) of Fig. 5(c) (not shown) and plot it as E_{J1} dependence in Fig. 6(b). In this case, like in Fig. 6(a), I_T and I_C show cosine-like dependence. However, most importantly, their correlation is now opposite to that in Fig. 6(a). This is consistent with the expectation that the state $\alpha|00\rangle + \beta|11\rangle$ is created.

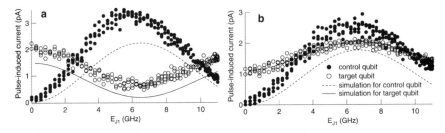

Figure 6. Pulse-induced current as a function of the Josephson energy of the control qubit under the pulse sequences shown in (a) Fig. 5(c) (i) and (b) Fig. 5(c) (ii). The lengths of the pulses in Fig. 5(c) (ii) are Δt_3=88 ps, Δt_4=264 ps, Δt_5=88 ps, Δt_{34}=88 ps and Δt_{45}=88 ps. The curves represent simulation obtained by calculating time evolution of the density matrix. In the calculation, we assumed a trapezoidal pulse shape with both rise and fall times equal to 40 ps, which is close to the real pulse shape. To take into account the effect of dephasing, all the off-diagonal terms of the density matrix are set to zero before applying the operation pulse. This is a reasonable approximation because the dephasing time at an off-degeneracy point is reported to be a few hundred picoseconds,[18] which is comparable to the time needed for the input preparation for the present experiment. We did not take into account the energy relaxation, which is known to be much slower.

The above data shows that we have succeeded with the conditional gate operation. However, to understand more quantitatively, we compare the data with simulation data obtained by numerically calculating the time evolution of the density matrix. The results of the simulation are shown as curves in Fig. 6. Here, we stress that no fitting parameters are used in the calculation.

First we consider the target qubit. Apart from the offset in Fig. 6(a), the simulated curves agree well with the experiment, suggesting that the oscillation amplitude of the measured I_T is reasonable. On the other hand, we have some discrepancy in I_C. We attribute this discrepancy to the unknown current channel in our present read-out scheme. As long as the JQP process is considered, the pulse-induced current should not be able to exceed $2e/T_r = 2.5$ pA, but in reality it does. This means that the pulse-induced current has an extra component that does not originate from the JQP process. We do not yet know the origin of this current. It may be other processes involving higher-order Cooper-pair tunnelling. The magnitude of this current probably depends on the Josephson energy (but does not depend strongly on the pulse length) and produces the E_{J1}-dependent deviation between the simulated and measured curves. In the target qubit, the similar current channel simply gives a constant offset in Fig. 6 as E_{J2} is fixed and does not affect the overall E_{J1}-dependence. Although quantitative analysis for I_C is difficult at present, the simulation suggests that the oscillation amplitude of the measured I_T is reasonable, while that of I_C is enhanced by this extrinsic factor originating from the imperfection of our read-out scheme.

In conclusion, we controlled our two-qubit solid-state circuit by applying a sequence of pulses and demonstrated the conditional gate operation. Although in the present experiment we paid attention only to the amplitude of the quantum state, phase evolution during the gate operation should also be examined for the realization of the quantum C-NOT gate (probably with additional phase factors), which is a constituent of the universal gate.

4. Single-shot readout of the Josephson charge qubit

In the experiments described above, individual probabilities of each qubit averaged over all states of the other qubit were measured. To directly measure multi-qubit states, one must be able to readout each qubit after every single-shot coherent state manipulation. The single-shot readout is of great importance, for instance, for quantum state tomography, quantum state teleportation, quantum cryptography.[17] Without the single-shot readout, algorithms that give non-unique solutions can not be utilized.

To readout single quantum states of the Josephson qubits (in particular, flux qubits) through the phase degree of freedom, a few circuits, measuring switching event from the supercurrent state to the finite-voltage state were implemented.[2-5] In charge type of qubits, it is straightforward to measure a charge quantum instead of the flux quantum.[19] For the single-shot charge readout, a radio-frequency single-electron transistor[20] electrostatically coupled to the qubit was proposed as a detector of the charge states.[21, 22] Although this approach works in principle,[23, 24] the single-shot readout has not yet been realized. In this work, we demonstrate an operation and study mechanism of novel readout scheme that allows to perform highly efficient single-shot measurements, with suppressed back-action of the measurement circuit on the qubit. An efficiency of the readout of $|1\rangle$ and $|0\rangle$ charge states are found to be 87% and 93%, respectively.

A scanning-electron micrograph of our circuit is shown in Fig. 5(a) The device consists of a charge qubit (3) and a readout circuit. The qubit is a Cooper-pair box (with its effective capacitance to the ground $C_b \approx 600$ aF) coupled to a reservoir through a Josephson junction with the Josephson energy $E_J \approx 20$ μeV. The reservoir is a big island with about 0.1 nF capacitance to the ground plane and galvanically isolated from the external environment. The qubit states are coherently controlled by a non-adiabatic control pulse, yielding a superposed state of $|0\rangle$ and $|1\rangle$. The readout part includes an electrometer which is a conventional low-frequency single-electron transistor (SET) ($C_s \approx 1000$ aF) and a charge trap ($C_t \approx 1000$ aF) placed between the qubit and the SET. The trap is connected to the box through a highly resistive tunnel junction ($R_t \approx 100$ MΩ) and coupled to the SET with a capacitance $C_{st} \approx 100$ aF. The use of the trap enables us to separate in time the coherent state manipulation and readout processes. Moreover, the qubit becomes effectively decoupled from the SET

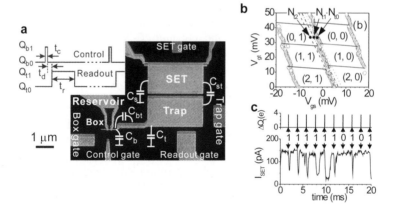

Figure 7. (a) Scanning electron micrograph of the device. The aluminum structure is deposited on top of a thin Si₃N₄ insulating layer (0.4 μm) above a gold ground plane. The device consists of a Cooper pair box, a reservoir, a trap and a measurement SET. The dc ("box" and "trap") and pulse (control and readout) gates control potentials of the islands. The inset schematically represents the pulse operation. (b) Stability diagram of the SET coupled to the trap. Open circles mark positions of the SET quasiparticle current peaks on V_{gs} - V_{gt} plane (dc gate voltages of the SET and the trap). Pairs of numbers (N_t, N_s) designate the trap - SET ground state charge configuration in each cell bounded by the SET peaks and black solid lines. Dashed, dashed-dotted and dotted lines indicate positions of the SET peaks for 0, 1 or 2 additional electrons in the trap, respectively. (c) A typical time-trace of the SET current (lower panel) together with the readout pulse sequence (upper panel). Digits 1 and 0 mark readout bits. Negative switches on the lower curve correspond to the detected charge of the trap.

(the qubit relaxation rate induced by the SET voltage noise is suppressed by a factor of $(C_{bt} C_{st}/C_t C_b)^2 \approx 3 \times 10^{-5}$, where $C_{bt} \approx 30$ aF). The coupling strength can be made even weaker, if the unwanted box-to-trap capacitance C_{bt} is further decreased.

The operation of the circuit can be described in the following way. During the qubit manipulation, the trap is kept unbiased prohibiting charge relaxation to the trap. Once the control pulse is terminated, the readout pulse (see the inset of Fig. 7(a)) is applied to the trap. The length and the amplitude of the readout pulse are adjusted so that if there is an extra Cooper pair in the box after termination of the control pulse, it escapes to the trap through a quasiparticle tunneling with a high probability. After the charge is trapped, it remains in the trap for a long time (a reverse trap-to-box charge relaxation is suppressed due to the superconducting energy gap 2Δ) and is measured by a low-frequency SET.

The Hamiltonian of the two-level system of the qubit in the charge basis $|0\rangle$ and $|1\rangle$ (without and with an extra Cooper pair) is $H = U_b(0, Q_b) |0\rangle\langle 0| + U_b(2, Q_b) |1\rangle\langle 1| - 1/2\, E_J (|0\rangle\langle 1| + |1\rangle\langle 0|)$ (we define an electrostatic energy of island k as $U_k(N_k, Q_k) = (N_k e - Q_k)^2/2C_k$, where k is either b or t indicating box or trap island, respectively, N_k is an excess electron number, and Q_k is a gate induced charge in the island. Starting at Q_{b0} ($Q_{b0} < Q_{b1}$), where $\Delta E \gg E_J$ ($\Delta E = U_b(2, Q_{b0}) - U_b(0, Q_{b0})$) we let the system relax to the ground state, which is nearly pure charge state $|0\rangle$. Then we instantly change the eigenbasis for a time t_c by applying a rectangular control pulse, which brings the system to Q_{b1}. If Q_{b1} is a degeneracy point ($\Delta E = 0$), the final state of the control pulse manipulation is $|0\rangle \cos \omega_J t_c/2 + |1\rangle \sin \omega_J t_c/2$ ($\omega_J = E_J/\hbar$), therefore, after the pulse termination, the state $|1\rangle$ is realized with a probability of $\cos^2 \omega_J t_c/2$.

Fig. 7(b) shows an experimentally measured stability diagram: SET current peak positions as a function of trap and box gate voltages of the SET and the trap. By setting the box and trap gates to one of the points N_{t0}, N_{t1} or N_{t2}, we can detect if the trap has 0, 1 or 2 additional electrons. In our measurements, the SET is usually adjusted to N_{t0}. To readout the qubit, the trap is biased by the readout pulse of typical length $t_r = 300$ ns and amplitude $\Delta Q_t = 3.5e$ ($\Delta Q_t = Q_{t1} - Q_{t0}$), applied to the readout gate, letting an extra Cooper pair of the state $|1\rangle$ escape to the trap through a quasiparticle tunneling and switching off the SET current (the SET peak position is shifted to the position of the dashed line).

The curve in the upper panel of Fig. 7(c) indicates a readout pulse sequence. The curve on the lower panel demonstrates a typical time-trace of the SET current. Negative switches on the curve of the lower panel coming synchronously with the readout pulses are counted as charge detection events. For the studied device, the lifetime of the trapped charge is typically about 300 µs, therefore, normally used repetition time of 2 ms is sufficient for practically complete trap resetting. We count the number of detected switches m, with the total number of shots n_{tot}.

An experimentally obtained probability of the charge detection $P = m/n_{tot}$ (normally, $n_{tot} = 327$ is used per one experimental data point) as a function of the control pulse length t_c and the amplitude ΔQ_b is shown as a two-dimensional plot in Fig. 8(a). We define the pulse with $\Delta Q_b = 0.84e$ ($\equiv Q_{bA}$) and $t_c = 120$ ps, when P reaches maximum, as a π-pulse. Fig. 8(b) demonstrates coherent oscillations as a function of t_c measured at Q_{bA}. As shown by the vertical arrowed line, the visibility here reaches 0.64, while the longest lasting oscillations shown in Fig. 8(c) are found at $\Delta Q_b = 0.75e$ ($\equiv Q_{bB}$), (the phase decoherence is expected to be the weakest at the degeneracy point). We believe that $Q_{bA} \neq Q_{bB}$ due to the control pulse distortion because of limited frequency bandwidths of the transmission lines and the pulse generator.

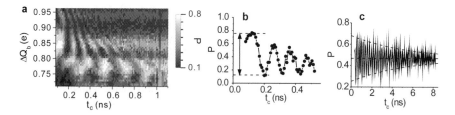

Figure 8. (a) Two-dimensional plot of coherent oscillations measured by averaging over many events of the single-shot measurements. (b) P versus t_c measured at $\Delta Q_b = 0.84e$ ($\equiv Q_{bA}$), where visibility is the highest. (c) P versus t_c measured at the degeneracy point, $\Delta Q_b = 0.75e$ ($\equiv Q_{bB}$), where the oscillations are the longest lasting. Dashed envelops correspond to the exponential decay with the decay time of 5.8 ns.

Figs. 9(a-b) demonstrate time relaxation dynamics of the qubit states. Fig. 9(a) shows a probability P to find an extra charge in the trap, when time delay t_d is introduced between the control π-pulse and the readout pulses. The exponential decay of P may be explained by tunneling to the reservoir (presumably via energetically feasible Cooper pair tunneling $(2, 0) \rightarrow (0, 0)$) because alternative quasiparticle relaxation to the trap through the high resistive junction is blocked by 2Δ when the trap is not biased. The relaxation rate to the reservoir found from the fitting (solid curve) is $\beta = (220\ \text{ns})^{-1}$. Fig. 9(b) shows relaxation dynamics of the state $|1\rangle$ as a function of the readout pulse length t_r ($t_d \approx 0$). This relaxation is mainly determined by quasiparticle decay to the trap with a rate α ($\alpha \gg \beta$). Additionally, the inset of Fig. 9(b) shows dynamics of $|0\rangle$-state relaxation ("dark" switches) to the trap. These "dark" switches can be presumably described by the process $(0, 0) \rightarrow (-2, 2)$, with a weak relaxation rate $\gamma = (4100\ \text{ns})^{-1}$ derived from fitting the data by $1 - \exp(-\gamma t_r)$ (solid curve).

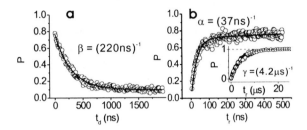

Figure 9. (a) P as a function of delay between the control π-pulse and the readout pulse t_d. A solid curve is a fitting by an exponent with a decay rate $\beta = (220 \text{ ns})^{-1}$. (b) A probability of $|1\rangle$-states detection created by the π-pulse as a function of the readout pulse length t_r. A solid curve is a result of fitting $m(t)$ using Eq. (4) normalized by n_{tot} with fitting parameters n_0^* and α ($n_0^*/n_{tot} = 0.87$ and $\alpha = (37 \text{ ns})^{-1}$). The inset shows probability without π-pulses. P is fitted by $1 - \exp[-\gamma t_r]$ with $\gamma = (4.1 \, \mu\text{s})^{-1}$.

Let us consider the relaxation dynamics in more details. The number of excited states, n^*, decreases within the time interval $[t, t + dt]$ as $dn^*(t) = -\alpha \, n^*(t) \, dt - \beta \, n^*(t) \, dt$. The number of states in $(0, 0)$ configuration, $n(t)$, changes, in turn, as $dn(t) = -\gamma \, n(t) \, dt + \beta \, n^*(t) \, dt$. We may also write an expression for the number of events in which the trap is found to be charged: $dm(t) = \alpha \, n^*(t) \, dt + \gamma \, n(t) \, dt$. Solving these equations with the initial conditions $n^*(0) = n_0^*$, $n(0) = n_{tot} - n_0^*$ and $m(0) = 0$ we find

$$m(t) = n_{tot}\left[1 - \exp(-\gamma t)\right] + n_0^* \frac{\alpha - \gamma}{\alpha + \beta - \gamma}\left[\exp(-\gamma t) - \exp(-(\alpha + \beta)t)\right]. \tag{4}$$

We fit the data of Fig. 9(b) by a curve of $P = m(t)/n_{tot}$ with $m(t)$ taken from Eq. (4) with two fitting parameters α and n_0^*. The fitting gives $\alpha = (37 \text{ ns})^{-1}$ and $n_0^*/n_{tot} = 0.84$, implying that the efficiency of $|0\rangle$-to-$|1\rangle$ conversion by the control π-pulse is 84%.

If our readout pulse length $t = t_r$ satisfies the conditions $\gamma \ll t_r^{-1} \ll \alpha + \beta$, then Eq. (4) can be simplified to

$$m(t_r) \approx n_0^* \frac{\alpha}{\alpha + \beta} + \gamma t_r\left[n_{tot} - n_0^* \frac{\alpha}{\alpha + \beta}\right]. \tag{5}$$

Using Eq. (5), one may estimate an efficiency of the single-shot readout. We introduce a probability $P_y(x)$ of finding the trap charged ($y = 1$) or uncharged ($y = 0$), when the qubit is in $|x\rangle$-state (x is either 0 or 1). According to the definition of $P_y(x)$, $P_0(0) + P_1(0) = 1$ and $P_0(1) + P_1(1) = 1$. The total number of detected events expressed in terms of these probabilities may be written as $m = n_0^* P_1(1) + (n_{tot} - n_0^*) P_1(0)$. Comparing the latter expression with Eq. (5) we find

$$P_1(1) = \frac{\alpha + \beta \eta_r}{\alpha + \beta},$$
(6a)

$$P_1(0) = \eta_r.$$
(6b)

Confirming that our readout pulse length $t_r = 300$ ns fulfills the necessary condition for Eq. (5), $\gamma \ll t_r^{-1} \ll \alpha + \beta$, we directly find from Eqs. (6) that the probability of detection of the state $|1\rangle$ is $P_1(1) = 0.87$ and the state $|0\rangle$ is $P_0(0) = 0.93$ ($P_1(0) = 0.07$). The readout efficiency can be further improved by optimizing the relaxation rates. The derived probabilities are consistent with the mean probability of the oscillations at the degeneracy point, $\langle P \rangle = [\langle n_0^* \rangle P_1(1) + (n_{tot} - \langle n_0^* \rangle) P_1(0)]/n_{tot} = 0.47$, where $\langle n_0^*/n_{tot} \rangle = 0.5$.

5. Energy Relaxation of the Josephson charge qubit

Apart from application of the single-shot readout for quantum computation, it is a good tool to study physics of the charge qubits. Here, we study decoherence of the charge qubits with help of the single-shot readout.

Decoherence of small Josephson circuits has been studied in a number of theoretical papers (see for example Refs. [25], [26]). Recently, a few experiments have measured in which decoherence of Josephson charge qubits for some special cases. For instance, in charge echo experiments, dephasing of the Josephson charge qubit has been measured far from the charge degeneracy point in Ref. [15].

Also, relaxation of excited states of the charge qubits off the degeneracy point has been measured in Ref. [27]. Dephasing and energy relaxation were studied at the degeneracy point of the qubit with charging-to-Josephson-energy ratio about unity, for which charge noise may be also important in Ref. [2]. However, to understand the origin of the decoherence, a systematic study of the qubit is of great importance.

In this work, we systematically measure and analyze the qubit energy relaxation in a wide range of the qubit parameters. It was found that the energy relaxation is caused by a quantum noise with approximately linear frequency dependence. We propose a model in which the quantum noise is caused by charge fluctuators, which give both the classical $1/f$ noise and the quantum f noise.

The sample schematically shown in Fig. 10(a) consists of a qubit and a readout part.[8] The qubit is a Cooper pair box connected to a reservoir through a tunnel junction of SQUID geometry with Josephson energy E_J controlled by an external magnetic field. The readout part contains a charge trap island and an electrometer – a single electron transistor (SET). The trap is connected to the box through a highly resistive tunnel junction. To read out the box charge state the trap is biased by a readout pulse applied to the readout gate, so that if the box is in the excited state, an extra Cooper pair tunnels in a sequential two-quasiparticle process and then measured by the SET. Note that the quantum states are not destroyed by the measurement circuit until the readout pulse is applied, as the SET is effectively decoupled from the qubit. We designate mutual and self capacitances of the islands by C_{ij} and C_i, where i and j are characters b, t, s denoting the box, trap and SET islands, respectively. The capacitance values are the following: $C_b \approx 0.6$ fF, $C_t \approx C_s \approx 1$ fF, $C_{bt} \approx 0.03$ fF, $C_{st} \approx 0.1$ fF. The charging energy of the box is $E_c \approx e^2/2C_b = 130$ μeV. The reservoir is a big island galvanically isolated from the external environment with its capacitance to the ground plane of the order of 0.1 nF. The SET is usually biased to the Josephson quasiparticle cycle peak[14] with the current of about 100 pA in maximum.

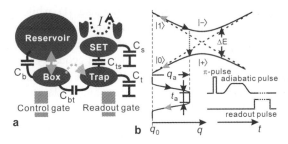

Figure 10. (a) Schematic device representation. (b) State manipulation diagram for measurements of the relaxations. The inset shows operation pulse sequence.

In the charge basis $|0\rangle$ and $|1\rangle$ (without and with an extra Cooper pair in the box) the Hamiltonian of the box can be written as

$$H = \Delta E/2 \,[\sigma_z \cos\theta - \sigma_x \sin\theta], \tag{7}$$

where σ_z, σ_x are the Pauli's matrices, $\Delta E = (U^2 + E_J^2)^{1/2}$, $U = 2eq/C_b$ is the electrostatic energy difference between the two states, q is the gate induced charge with "zero" shifted to the degeneracy point and $\theta = \arctan(E_J/U)$. Eigenstates of the two-level system are $|+\rangle = \cos\theta/2\,|0\rangle + \sin\theta/2\,|1\rangle$ and $|-\rangle = -$

sin θ/2 |0⟩ + cos θ/2 |1⟩ with the corresponding eigenenergies -$\Delta E/2$ and $\Delta E/2$. Fig. 10(b) shows schematically the energy diagram of the qubit as a function of q. Solid and dashed lines represent eigenenergies and electrostatic energies, respectively. We usually adjust the qubit to a position of q_0 where $U \gg E_J$ and eigenenergies are nearly pure charge states |+⟩ ≈ |0⟩ and |-⟩ ≈ |1⟩. Starting from the ground state |0⟩, we apply a rectangular (non-adiabatic) pulse bringing the system to the degeneracy point (θ = π/2) for the time t_p, where the state freely evolves as cos $E_J/2\hbar$ t |0⟩ + sin $E_J/2\hbar$ t |1⟩. When the pulse length $t_p = \pi\hbar/E_J$ (π-pulse), the evolution yields state |1⟩.

To measure energy relaxation dynamics of the excited state |-⟩ we use a combination of the π-pulse and an adiabatic pulse (a pulse with slow rise and fall times satisfying the condition of $\hbar|d\Delta E/dt| \ll E_J^2$). The manipulation procedure, schematically shown in Fig. 10(b), includes three sequential steps: First, the π-pulse is applied to the box to prepare the excited state |1⟩. Second, an adiabatic pulse is applied to the box, so that its rise front shifts the system along the excited state |-⟩ to a point $q = q_0 + q_a$ and holds the system in these conditions for a time ta, where relaxation from the excited state |-⟩ to the ground state |+⟩ may occur with a finite probability dependent on ta. Third, the fall of the adiabatic pulse converts the excited state |-⟩ to the charge state |1⟩, while the ground state |+⟩ is converted to the state |0⟩. One can study relaxation dynamics at a desired value of q by measuring probabilities of states |1⟩ as a function of time t_a.

We study two samples (I and II) of an identical geometry at a temperature of 50 mK. Probability P of the excited state |1⟩ is determined by repeating nominally identical quantum state manipulations and readouts.[8] The inset of Fig. 11 (a) exemplifies a typical decay of P of excited states measured as a function of ta at q = - 0.36 e of sample I. We derive the energy relaxation rate Γ_1 fitting the data by an exponent $A \exp[-\Gamma_1 t_a]$ + B with three fitting parameters A, B and Γ_1. The amplitude A depends on the efficiency of each step of the state manipulations with t_a = 0 and is a constant at each fixed q, because all other parameters (pulse lengths, delay times and pulse amplitudes) are kept unchanged within one data point measurements. A small finite value of B is a consequence of "dark" switches in our circuit. Note that B is always small ($B \ll 1$) independently of q indicating that relaxation is much stronger than excitation.

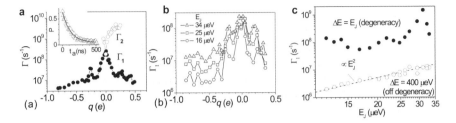

Figure 11. (a) Energy relaxation rate Γ_1 (closed circles) versus gate induced charge q of sample I (E_J = 21 μeV) and phase decoherence rate Γ_2 (open circles). The inset exemplifies a decay of probability P to detect charge in the trap as a function of the adiabatic pulse length t_a, measured at q = -0.36 e. Γ_1 is derived from exponential fit of the decay. (b) Γ_1 measured in sample II for three different Josephson energies: E_J = 16 μeV (dots), E_J = 25 μeV (squares) and E_J = 34 μeV (triangles). (c) Relaxation rate Γ_1 as a function of E_J measured at the degeneracy point (closed circles) and off the degeneracy point (open circles) at ΔE = 400 μeV for sample I.

Figures 11(a), (b) show Γ_1 as a function of q. Fig. 11(a) demonstrates Γ_1 (closed circles) measured at E_J = 20 μeV of sample I. For this sample, we also derive the phase decoherence rate Γ_2 (open circles) in a limited range of q from the decay of the coherent oscillations. Fig. 11(b) shows Γ_1 dependences measured for three different Josephson energies: E_J = 16 μeV (dots), E_J = 25 μeV (squares) and E_J = 34 μeV (triangles) of sample II. E_J dependences of the relaxation rate Γ_1 for the sample II at fixed adiabatic pulse amplitudes are shown in Fig. 11(c). Γ_1 at q = -0.8 e (ΔE = 400 μeV) is presented by open circles, while Γ_1 at q_a = 0 (degeneracy point, U = 0) is presented by closed circles.

For the charge qubit, fluctuations coupled to the charge degree of freedom are, presumably, the main origin of decoherence. If the energy fluctuations of the qubit due to the charge noise are characterized by the spectral density $S_E(\omega)$, the relaxation rate is given by (see, for example, Ref. [28])

$$\Gamma_1 = \pi S_E(\omega)/2\hbar^2 \sin^2 \theta. \qquad (8)$$

The overall behavior of Γ_1 in Figs. 11(a), (b) (Γ_1 decreases off the degeneracy point) is described by $\sin^2\theta = E_J^2/(E_J^2 + U^2)$, characterizing coupling of the qubit to the reservoir. When $\Delta E \gg E_J$, Eq. (8) can be rewritten in the form $\Gamma_1 \approx \pi S_E(\omega \approx U/\hbar)(E_J/U)^2/\pi\hbar^2$. The experimentally measured Γ_1 at ΔE = 400 μeV (open circles in Fig. 11(c)) has clear E_J^2 dependence (dashed curve). At the degeneracy point, $\Gamma_1 = \pi S_E(\omega = E_J/\hbar)/2\hbar^2$ (closed circles in Fig. 11(c)) directly reproduces frequency dependence of S_E. We argue that in our well electrically isolated device, the relaxation can not be explained by spontaneous emission to the remote environment connected to the qubit through electrical leads of the gates and the readout circuit. We estimate the relaxation rate at the degeneracy point $q = 0$ ($\theta = \pi/2$), where the qubit is most sensitive to the charge noise. The relaxation induced by coupling to the "ohmic" environment through the control gates can be estimated using Eq. (8) with the real part of the gate impedance Re $Z_g \approx 100\ \Omega$ at $\hbar\omega \gg kT$ and the qubit coupling strength to the gates κ_g. $\kappa_g = C_{cg}/C_b = 2\times10^{-3}$ for the control gate with the gate-to-box capacitance $C_{cg} \approx 1\text{aF}$ and $\kappa'_g \approx C_{rg}\ C_{bt}/C_b C_t = 5\times10^{-3}$ for the readout gate with the gate-to-trap capacitance $C_{rg} \approx 10$ aF. Substituting the noise spectrum $S_E = (2\ e^2\kappa^2/\pi)\ 2\text{Re}\ Z_g\ \hbar_\omega$, we obtain $\Gamma_1 \approx 3\times10^4\ \text{s}^{-1}$ for $\kappa_g = 5\times10^{-3}$ for the frequency $\omega = 2\pi \times 5$ GHz. Similarly, we estimate an upper limit of the noise induced by the measurement SET taking the SET-to-box coupling strength $\kappa_{sb} \approx C_{st}\ C_{tb}/C_t\ C_b = 5\times10^{-3}$ and the real part of the SET impedance $Z_{SET} = r/(1 + i\omega r C_S)$ ($r \approx 100$ kΩ is the effective resistance of two SET junctions in parallel). For $\omega > 2\pi \times 5$ GHz, Re(Z_{SET}) $\approx (\omega^2\ r\ C_s^2)^{-1}$ and $Re(Z_{SET}) < 3$ kΩ, $\Gamma1 < 10^6\ \text{s}^{-1}$. Decoherence induced by the noise of the SET measurement current produces classical noise ($S_E(\omega) = S_E(-\omega)$), resulting in equally populated excited and ground states, rather than relaxation. Nevertheless, it is worth showing that the measurement circuit is not the main source of decoherence of the qubit. For the SET current $I_s \approx 100$ pA, electron tunneling rate $I_s/e \sim 10^9\ \text{s}^{-1}$, which, in our case of $\omega > 2\pi\ I_s/e$, induces $S_E \approx 2/\pi\ (e^2/C_s)^2\ \kappa_{sb}^2\ (I_s/e)/\omega^2$. The corresponding relaxation rate at $\omega > 2\pi \times 5$ GHz is $\Gamma_1 < 4\times10^6\ \text{s}^{-1}$.

It is believed that one of the main origins of the low frequency noise for the nano-scale charge devices is charge fluctuations in the substrate or in the junction with a spectrum given by

$$S_q(\omega) = \alpha/(2|\omega|).$$

(9)

Note that here the $1/f$ noise is defined for both positive and negative frequencies. The parameter α has been measured for a device with essentially same geometry and fabrication process as ours, and has been found to be $\alpha = (1.3\times10^{-3}\ e)^2$. This value is very typical for the $1/f$ noise measured in other works.[29, 30] It was shown in charge echo experiments that the $1/f$ noise extrapolated to the gigahertz range reasonably (with an accuracy of a factor of 2) describes the dephasing of the charge qubit. In the case of the Gaussian noise approximation from many fluctuators weakly coupled to the qubit, the coherent oscillations dephase as exp[-φ], as the random phase

$$\varphi(t_c) \approx \frac{\cos^2\theta}{\hbar^2} \int_{\omega_0}^{\infty} S_E(\omega) \left(\frac{2\sin(\omega t_c/2)}{\omega}\right)^2 d\omega$$

(10)

is accumulated at frequencies $|\omega| \leq 2\pi/t_c$, where $S_E(\omega) = (4E_c/e)2S_q(\omega)$. The phase decoherence time $T_2 = \Gamma_2^{-1}$ is defined as $\varphi(T_2) = 1$. Taking the noise spectrum from Eq. (9), we can find the parameter $\alpha = \eta\hbar e/(E_cT_2\cos\theta)$, where η is a numeric coefficient with an extremely weak dependence on the lower cutoff frequency ω_0. Taking $\omega_0 = \pi/\tau_m$ with a typical measurement time of one data point $\tau_m = 1$ s, we find $\eta \approx 0.053$. Using the experimentally measured T_2 for the sample I, we obtain $\alpha \approx (1.3\times10^{-3}e)^2$.

Fig. 12(a) summarizes reduced noise spectra $S_E/\hbar^2 = 2\Gamma_1/(\pi\hbar^2\sin^2\theta)$ derived from dependences of Γ_1 according to Eq. (9). $S_E/\hbar^2$ derived from Γ_1 - q dependences in sample I is plotted by closed circles, while $S_E/\hbar^2$ for sample II with different E_J is plotted by open circles. In addition, $S_E/\hbar^2 = 2\Gamma_1/(\pi\hbar^2)$ measured at the degeneracy point of sample II is plotted by open triangles. We also show the $S_E/\hbar^2$ for the $1/f$ noise with $\alpha = (1.3\times10^{-3}\ e)^2$ (see Eq. (10)) by the dashed line. $S_E/\hbar^2$ in Fig. 12(a) exhibits rise with increasing ω. The dashed-dotted line exemplifies linear dependence (as in the case of an "ohmic" environment), which we present in the form of $2R\hbar\omega e^2/\hbar^2$ with $R = 6\ \Omega$. The actual rise of the experimental data is not monotonic but has some resonance-like peaks, for instance, at 7 and 30 GHz. This probably reflects coupling to some resonances, which can be two-level oscillators or simply geometrical resonances in the sample package. Γ_1 approaches the $1/f$ noise (dashed line) at low frequencies. The crossover frequency of the $1/f$ and f curves is $\omega_c = 2\pi\times2.6$ GHz, which formally corresponds to the effective temperature $T_C = \hbar\omega_c/k = 120$ mK.

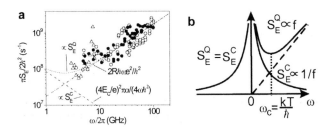

Figure 12. (a) Noise $S_E/\hbar^2$ derived from Γ_1 - q dependences of the sample I (closed circles) and sample II (open circles). Open triangles show Γ_1 dependences of sample II at the degeneracy point. A dashed-dotted line represents a linear rise $2e^2R\hbar\omega$ (f noise) with $R = 6\ \Omega$. A dashed line is the $1/f$ and f noises from Eq. (9) with $\alpha \approx (1.3\times10^{-3}\ e)^2$ derived from Γ_2 measurements. Additional dotted line presents a sum of (b) A schematic representation of the classical $S^C_E(\omega)$ and quantum $S^Q_E(\omega)$ noise behavior. At frequencies $\omega < kT/\hbar$, $S^Q_E(\omega) \approx S^C_E(\omega)$ and behaves like a $1/f$ noise. At frequencies $\omega > kT/\hbar$, $S^Q_E(\omega) > S^C_E(\omega)$ and the quantum noise is proportional to ω.

We found that T_C is very close to the expected electron temperature. This fact suggests us to propose the following model which may explain the measured noise spectra. We assume that the main noise originates from charge fluctuations of a bath of fluctuators at effective temperature T_C weakly coupled to the qubit. Pure dephasing should be caused by fluctuators in thermal equilibrium, which produce the classical noise S^C_E, while the energy relaxation is caused by the quantum noise S^Q_E ($S^Q_E(|\omega|) > S^Q_E(-|\omega|)$). Namely, the "hot" qubit ($\Delta E > kT_C$) may only release the excess energy ΔE to the "cold" fluctuator bath. On a microscopic level, this means that the excess energy ΔE may activate fluctuators of the "cold" bath, that are quiet in equilibrium (do not contribute to S^C_E). Fig. 12(b) schematically represents the noise behavior extended to a negative frequency range. S^Q_E almost coincides with S^C_E at low frequencies $\hbar\omega < kT_C$ ("hot" fluctuators), because the qubit has an equal chance to emit and to absorb energy. As we discussed above, S^C_E is expected to have $1/f$ dependence, while S^Q_E is roughly proportional to f according to our experimental data, and consistent with the "ohmic" picture.

Within this model one can predict the classical noise if the energy relaxation is known. For instance, in the discussed case of $1/f$ and f noises, if the absorption f-part is $S^Q_q = \beta\hbar\omega$, at $\omega \gg \omega_c$ ($S_E = (4E_c/e)^2 S_q$), the $1/f$ classical noise can be found from the condition $\alpha/2\omega_c = \beta\hbar\omega_c$, where β is a parameter independent of temperature. The corresponding quantum noise is

$$S^Q_q = \frac{\beta(kT_{eff})^2}{\hbar|\omega|^2} + \beta\hbar\omega\Theta(\omega) \tag{11}$$

where the first term is the classical $1/f$ noise S^C_E and the second term is the f quantum noise S^Q_E, $\Theta(\omega)$ is the step function. As it follows from Eq. (11), in this model $\alpha = \beta (kT_C)2/\hbar$ is expected to be proportional to T_C^2 because of the scale invariance. That is, if the density of states of the fluctuators is a function of two independent energies (for example, excitation energies of electrons in a double-well potential) then the $1/f$ noise is proportional to T_C^2. $1/f$ charge noise has been measured in a wide temperature range using high charging energy SET.[31] Indeed, T^2-dependence was observed in a few samples.

Finally, we conclude that the noise spectra, shown in Fig. 12(a) can be caused by the bath of fluctuators. At low frequencies, the fluctuators should give rise to the $1/f$ classical noise spectrum $S^C_E(\omega)$, yielding the qubit dephasing. However, at high frequencies, $\omega > kT_C/\hbar$, the qubit relaxation is caused by the quantum noise $S^Q_E(\omega)$ which may originate from the bath of fluctuators. The quantum noise tends to coincide with the classical noise $S^Q_E(\omega) = S^C_E(\omega)$ at $\omega < kT_C/\hbar$.

References

1. Y. Nakamura, Yu. A. Pashkin, J. S. Tsai, *Nature* **398**, 786 (1999).
2. D. Vion, A. Aassime, A. Cottet, P. Joyez, H. Pothier, C. Urbina, D. Esteve, M. H. Devoret, *Science* **296**, 886 (2002).
3. Y. Yu, S. Han, X. Chu, S.-I. Chu, Z. Wang, *Science* **296**, 889 (2002).
4. J. M. Martinis, S. Nam, J. Aumentado, C. Urbina, *Phys. Rev. Lett.* **89**, 117901 (2002).
5. I. Chiorescu, Y. Nakamura, C. J. P. M. Harmans, J. E. Mooij, *Science* **299**, 1869 (2003).
6. Yu. A. Pashkin, T. Yamamoto, O. Astafiev, Y. Nakamura, J. S. Tsai, *Nature* **421**, 823 (2003).
7. T. Yamamoto, Y. A. Pashkin, O. Astafiev, Y. Nakamura, J. S. Tsai, *Nature* **425**, 941 (2003).
8. O. Astafiev, Yu. A. Pashkin, T. Yamamoto, Y. Nakamura, J. S. Tsai, *Phys. Rev.* **B69** (*Rapid Communications*), 180507 (2004).
9. O. Astafiev, Yu. A. Pashkin, T. Yamamoto, Y. Nakamura, J. S. Tsai, *Phys. Rev. Lett.* **93**, 267007 (2004).
10. A. Shnirman, G. Schön, Z. Hermon, *Phys. Rev. Lett.* **79**, 2371 (1997).
11. D. V. Averin, *Solid State Commun.* **105**, 659 (1998).
12. V. Bouchiat, D. Vion, P. Joyez, D. Esteve, M. H. Devoret, *Physica Scripta* **T76**, 165 (1998).
13. H. Pothier, C. Lafarge, C. Urbina, D. Esteve, M. H. Devoret, *Europhys. Lett.* **17**, 249 (1992).
14. T. A. Fulton, P. L. Gammel, D. J. Bishop, L. N. Dunkleberger, G. J. Dolan, *Phys. Rev. Lett.* **63**, 1307 (1989).
15. Y. Nakamura, Yu. A. Pashkin, T. Yamamoto, J. S. Tsai, *Phys. Rev. Lett.* **88**, 047901 (2002).
16. J. R. Friedman, V. Patel, W. Chen, S. K. Toltygo, J. E. Lukens, *Nature* **406**, 43 (2000).
17. M. A. Nielsen, I. L. Chuang, *Quantum Computation and Quantum Information.* (Cambridge Univ. Press, Cambridge, 2000).
18. C. H. Bennett, H. J. Bernstein, S. Popescu, B. Schumacher, *Phys. Rev.* **A53**, 2046 (1996).
19. A. Shnirman, G. Schön, *Phys. Rev.* **B57**, 15400 (1998).
20. R. J. Schoelkopf, P. Wahlgren, A. A. Kozhevnikov, P. Delsing, D. E. Prober, *Science* **280**, 1238 (1998).
21. A. Aassime, G. Johansson, G. Wendin, R. J. Schoelkopf, P. Delsing, *Phys. Rev. Lett.* **86**, 3376 (2001).
22. G. Johansson, A. Käck, G. Wendin, *Phys. Rev. Lett.* **88**, 046802 (2002).
23. T. Duty, D. Gunnarsson, K. Bladh, P. Delsing, *Phys. Rev.* **B69**, 140503 (2004).

24. K. W. Lehnert, K. Bladh, L. F. Spietz, D. Gunnarsson, D. I. Schuster, P. Delsing, R. J. Schoelkopf, *Phys. Rev. Lett.* **90**, 027002 (2003)
25. Y. Makhlin, G. Schön, A. Shnirman, *Rev. Mod. Phys.* **73**, 357 (2001).
26. D. V. Averin, *Fortshr. Phys.* **48**, 1055 (2000).
27. K. W. Lehnert, K. Bladh, L. F. Spietz, D. Gunnarsson, D. I. Schuster, P. Delsing, T.R. J. Schoelkopf, *Phys. Rev. Lett.* **90**, 027002 (2003).
28. R. J. Schoelkopf, A. A. Clerk, S. M. Girvin, K. W. Lehert, M. Devotret, pp. 175-203, in Quantum noise in mesoscopic physics. Edited by Y. V. Nazarov, *Kluwer Academic Publishers*.
29. G. Zimmerli, T. M. Eiles, R. L. Kautz, J. M. Martinis, *Appl. Phys. Lett.* **61**, 237 (1992).
30. S. M. Verbrugh, M. L. Benhamadi, E. H. Visscher, J. E. Mooij, *J. Appl. Phys.* **78**, 2830 (1995).
31. Yu. A. Pashkin, Y. Nakamura, J. S. Tsai, *Appl. Phys. Lett.* **76**, 2256 (2000).

PULSED ENDOR-BASED QUANTUM INFORMATION PROCESSING

ROBABEH RAHIMI[1], KAZUNOBU SATO[2], KOU FURUKAWA[3],
KAZUO TOYOTA[2], DAISUKE SHIOMI[2] , TOSHIHIRO NAKAMURA[2], MASAHIRO
KITAGAWA[1], AND TAKEJI TAKUI[2†]

[1]*Department of System Innovation, Graduate School of Engineering Science,
Osaka University, Toyonaka, 560-8531, Japan*

[2]*Departments of Chemistry and Materials Sciences, Graduate School of Science,
Osaka City University, Sugimoto, Sumiyoshi-ku, Osaka 558-8585, Japan*

[3]*Institute for Molecular Science, Okazaki, 444-8585, Japan*

Pulsed Electron Nuclear DOuble Resonance (pulsed ENDOR) has been studied for realization of quantum algorithms, emphasizing the implementation of organic molecular entities with an electron spin and a nuclear spin for quantum information processing. The scheme has been examined in terms of quantum information processing. Particularly, superdense coding has been implemented from the experimental side and the preliminary results are represented as theoretical expectations.

1. Introduction

A quantum computer which can give all the advantages of quantum information processing (QIP) and quantum computation (QC) has been the focus of the contemporary issues in quantum science and several physical systems have been introduced, being examined to catch the idea from the experimental side.

Liquid-state nuclear magnetic resonance (NMR) spectroscopy has been used widely for implementations of even considerably complicated quantum non-local algorithms and the experimental outcomes represent the capability of NMR as the physical systems for QIP and QC [1-7].

NMR spectroscopy can be regarded such a physical system as represents some very important advantages for the realization of a quantum computer; The existence of nuclear spins with long decoherence time is just a proper physical realization of a qubit as the spin manipulation which can be easily performed by introducing the radio frequency pulses with relevant resonance frequencies. In

---

† takui@sci.osaka-cu.ac.jp

this context, we have emphasized that in terms of spin manipulation QC physical systems involving nuclear spins are preferable and novel materials as part of the QC physical system should be searched.

Recently, we have been working with pulse-based Electron Nuclear DOuble Resonance (ENDOR) [8] as a physical system to realize a quantum computer by invoking molecular entities with both electron spins and nuclear spins in the solid state. Since the physical system involves nuclear spins in organic open-shell molecular entities, pulsed ENDOR based quantum computer retains the main advantages of NMR systems. Nevertheless, QC-ENDOR as a double magnetic resonance spectroscopy is much heavier experimental task compared with electron paramagnetic (spin) resonance (EPR/ESR), but QC-ENDOR rewards back the much more efforts by adding up the advantages of NMR and EPR. Also, molecule-based ENDOR can afford potential applications of a variety of entities to QIP and QC. An elaborate total design of the QC-ENDOR experimental setup should be associated with molecular design for open-shell entities.

2. ENDOR based quantum computer

For any physical system as a candidate for the realization of a quantum computer, there are some criteria, known as DiVincenzo criteria that should be met. Molecule-based ENDOR systems should also meet these criteria to be regarded as a reliable physical system for QIP and QC. In the ENDOR-based QIP and QC, molecular electron spins in addition to nuclear spins have been introduced as quantum bits. To our knowledge, in terms of existing open-shell molecular entities the feasibility of quantum operations in molecule-based ENDOR alone are limited because of the nature of nuclear-nuclear interactions appearing in the ENDOR spectra [9]. In this work, only some aspects related to the criteria are discussed.

With molecule-based ENDOR system, the situation for the required experimental conditions for preparation of the initial state for QIP and QC seems to be achievable. As a possible approach, the high spin polarization of the electron spin can be transferred to the nuclear spins by applying the relevant pulse sequences followed by proper waiting times.

In this study, it has turned out that considerably long decoherence times of the systems composed of nuclear spins and an electron spin ($S = 1/2$) are available for which quantum operations between the two spins are demonstrated. Properly designed molecular entities with long decoherence times for demonstrations of multi-qubit operations seem not to be out of reach, for which QC-ENDOR experiments are underway.

In molecule-based ENDOR, manipulation and processing on the qubits as well as the readout processing can be realized by introducing radio frequency pulses on the nuclear spins and/or the microwave frequency pulses on the electron spin (see Figure 1). In this work, we have been mainly engaged in two experimental tasks for the realization of a quantum computer by molecule-based ENDOR. One has been an attempt to prepare the experimental requirements for demonstrating a true entanglement between a molecular electron spin and a nuclear spin by the use of a simple organic radical in the corresponding host single crystal. High spin polarizations on both the electron and nuclear spins are essentially required to achieve the true entanglement between the two spins in the molecular frame.

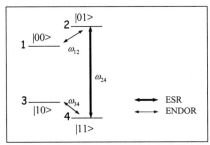

Figure 1. Energy levels and corresponding EPR/ESR and ENDOR resonance frequencies in the presence of a static magnetic field.

Investigation of the entanglement trough the spin Hamiltonian for the ENDOR system composed of only two electron and nuclear spins gives a necessary temperature of 0.8 K in a static magnetic field for the microwave transition frequency of 95 GHz, as given by the negativity criterion. Whereas, if the pulses can be applied for the transferring of the high spin polarization, the required temperature at the same magnetic field is nearly 5.1 K, which is well in reach with the current technology with a W-band (95 GHz) ENDOR spectrometer operating at liquid Helium temperature.

In this study, while the preparation of all the experimental requirements is in progress, the efforts still can be maintained on some other aspects of the research. Among those, one is materials challenge to design and synthesize stable open-shell molecular entities suitable for QIP/QC ENDOR experiments. Novel molecular open-shell systems with stable isotope labeled organic compounds for our purposes have also been designed and synthesized [9]. Also, the critical temperature can be tuned by invoking stable high-spin molecular entities.

Our efforts in this contribution have been made on the investigation of the credibility of the pulsed-ENDOR based QIP and QC to develop the necessary quantum gates and the entangling unitary operations. It is clear that in case of acquiring the former task with the achievement of the experimental conditions for the entanglement there would be no need to have an additional experimental processing to make the pseudo-pure state.

In order to check the credibility of the ENDOR physical system for QIP and QC and also to check the feasibility of the molecule-based QC-ENDOR with the current technology, implementation of superdense coding (SDC) has been revisited in our experiments. Pulsed ENDOR technique has been applied to a molecular electron- and nuclear-spin system, malonyl radical, $CH(COOH)$ in the single crystal of malonic acid, $CH_2(COOH)$ [10], in order to implement the SDC. In the malonyl radical system, a nucleus with a large hyperfine coupling, i.e., the α-proton has been selected.

3. Implementation of SDC by ENDOR

Superdense coding introduced by Bennett and Wiesner [11] is a non-local quantum algorithm in which two classical bits of information are transformed from Alice to Bob by sending a single qubit alone. The scheme is based on the fact that the entangled initial states are shared between two involved parties.

Alice encodes the qubit by applying a unitary transformation out of the four choices of {I, X, Y, Z} and then sends the qubit to Bob, who has been also initially given a qubit entangled to the Alices's one. After receiving the encoded qubit, he applies a measurement in the Bell basis on the both of the qubits and extracts the information on the Alices's choice. Therefore, a two-bit message is transferred by sending just a single qubit to one' partner.

Superdense coding has been implemented using some quantum physical systems including NMR [5]. In this contribution, we report on the implementation of the SDC by ENDOR. The main idea is just to test the ENDOR system for QIP rather than the implementation of superdense coding, giving a testing ground for QIP and QC to molecule-based ENDOR. As the latter one has been argued to be truly realizable in case of manipulating the entangled state and not just by a pseudo-entanglement. In our experiment, states have been prepared as pseudo-pure ones.

We also have used a single-crystal of malonic acid that incorporates malonyl radicals ($S = 1/2$) after X-ray irradiation at ambient temperature.

Existence of one α–proton with a large hyperfine coupling gives two spins as the required two qubits for superdense coding. The large hyperfine interaction is required to enable us to make a selective microwave excitation. Energy levels and the corresponding resonance frequencies of the radio and microwave frequencies are shown in Figure 1. Detection of the pseudo-entanglement with this system has been already reported [12] and in a similar approach we demonstrate the implementation of superdense coding by the use of pseudo-entangled states.

In order to get the necessary information on the resonance frequencies of the electron and nuclear spins, the pulsed EPR and ENDOR spectra have been measured in the single-crystal sample in the wide range of temperature. The representative spectra of the pulsed EPR and ENDOR are shown in Figures 2-3, respectively. In the pulsed ENDOR measurement, the radio frequency pulses have been applied on the sample in a range of 0-45 MHz with RF amplifiers of 300 W and 500 W.

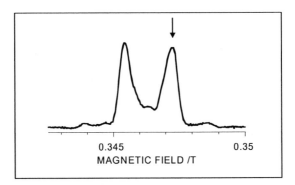

Figure 2: Pulsed EPR spectrum of malonyl radical in the single crystal. The arrow indicates the static magnetic field for the ENDOR measurement.

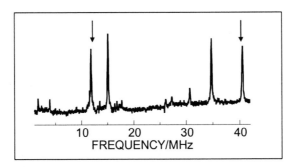

Figure 3: Pulsed ENDOR spectrum through measurement on the malonyl radical.

The pulse sequence that has been used for the implementation of SDC with ENDOR is given in Figure 4. There are three main parts, i.e., the preparation of the pseudo-pure states, manipulation and finally detection.

The outermost the left part labeled by "A" is for the preparation of the pseudo-pure state. Two pulses on electron and nuclear spins with additional waiting times in order to make the off diagonal terms of the density matrix vanishing are required to get the pseudo-pure state.

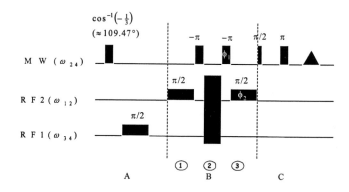

Figure 4: Pulse sequence for implementation of superdense coding by ENDOR.

The first two pulses in the central part of the sequence are for entangling and also the last two pulses are for the detection of the entanglement as reported by Ref. [12]. Two phases of Φ_1 and Φ_2 for the pulses in the detection part are required for clearing up the entangled states from the simple superposition states. The central part between the entangling and detecting the entanglement, which is labeled by B2, one of the qubits, nuclear spin in our experiment, is encoded by randomly applying one of the four pulses of {I, X, Y, Z}. The necessary pulses for encoding are described in Table 1.

Table 1: The unitary operation and corresponding pulse sequences for encoding.

	Initial state	Necessary operation	Required pulses	encoded state
U=I	$\frac{1}{\sqrt{2}}(\lvert 00\rangle+\lvert 11\rangle)$			$\frac{1}{\sqrt{2}}(\lvert 00\rangle+\lvert 11\rangle)$
U=X	$\frac{1}{\sqrt{2}}(\lvert 00\rangle+\lvert 11\rangle)$	$\exp(-i\pi I_x)$	$P_x^{34}(\pi)P_x^{12}(\pi)$	$\frac{1}{\sqrt{2}}(\lvert 01\rangle+\lvert 10\rangle)$
U=Y	$\frac{1}{\sqrt{2}}(\lvert 00\rangle+\lvert 11\rangle)$	$\exp(-i\pi I_y)$	$P_x^{34}(2\pi)P_x^{34}(\pi)P_x^{12}(\pi)$	$\frac{1}{\sqrt{2}}(\lvert 01\rangle-\lvert 10\rangle)$
U=Z	$\frac{1}{\sqrt{2}}(\lvert 00\rangle+\lvert 11\rangle)$	$\exp(-i\pi I_z)$	$P_x^{34}(2\pi)$	$\frac{1}{\sqrt{2}}(\lvert 00\rangle-\lvert 11\rangle)$

Finally, there are pulses for the detection by an electron spin echo signal. For the measurement part, the situation has been modified for some detection considerations. For the measurement part in this study we use the electron spin echo detection. The echo intensities have been detected for different angular dependencies of the pulses in the encoding part. As a result, there are four sets of angular dependencies for radiofrequency pulses which have been used for encoding, (see Table 2). Modified detection schemes besides the echo detection are available.

Table 2: Detection through angular dependence of the intensities of the electron spin echo.

	Angular dependent radiofrequency pulses for encoding	Detected angular dependent echo intensity
U=I		$\frac{1}{4}(-1+\cos[\phi_1-\phi_2])$
U=X	$\omega_x^{34}(\theta)\omega_x^{12}(\phi)$	$\frac{1}{16}(-3\cos\theta-\cos\phi+4\cos\frac{\theta}{2}\cos\frac{\phi}{2}\cos[\phi_1-\phi_2])$
U=Y	$\omega_x^{34}(2\pi+\theta)\omega_x^{12}(\phi)$	$\frac{1}{16}(-3\cos\theta-\cos\phi-4\cos\frac{\theta}{2}\cos\frac{\phi}{2}\cos[\phi_1-\phi_2])$
U=Z	$\omega_x^{34}(\theta)$	$\frac{1}{16}(-1-3\cos\theta+4\cos\frac{\theta}{2}\cos[\phi_1-\phi_2])$

4. Discussions

We have examined a pulsed Electron Nuclear DOuble Resonance (ENDOR) based approach to quantum information processing and quantum computation by invoking molecular electron and nuclear spins in the solid state. Pulse ENDOR technology retains the main advantages of NMR systems, and at the same times it seems to be useful and complementary to NMR technology in establishing entanglement status. We have noticed that time domain of electron spin manipulation technology is shortened and its present drawback is a matter of time resolution from the experimental side.

At first, our efforts have been made to develop entangling unitary operations in order to check the credibility of the molecule-based ENDOR QIP from both the theoretical and experimental sides. Thus, we have implemented superdence coding by acquiring the pseudo-pure states. The preliminary results are well representing the feasibility of the ENDOR quantum information processing.

Then, we are planning to examine the real quantum information processing and quantum computation by avoiding pseudo-pure states and introducing pure states in order to prepare the quantum entanglement between molecular electron spins and nuclear spins. It has been shown that the experimental requirements for achievement of spins with high polarization are not far beyond the current ENDOR spin-manipulation technology.

References

1. J. A. Jones, M. Mosca, and R. H. Hansen, *Nature* **393**, 344 (1998).
2. I. L. Chuang, L. M. K.Vandersypen, X. Zhou, D. W. Leung, and S. Lloyd, *Nature* **393**, 143 (1998).
3. M. A. Nielsen, E. Knill, and R. Laflamme, *Nature* **396** 52 (1998).
4. L. M. K. Vandersypen, M. Steffen, G. Breyta, C. S. Yannoni, M. H. Sherwood, and I. Chuang, *Nature* **414** 883 (2001).
5. X. Fang, X. Zhu, M. Feng, X. Mao, and F. Du, *Phys. Rev. A* **61**, 022307 (2000).
6. D. G. Cory, A. F. Fahmy, and T. F. Havel, *Proc. Natl. Acad. Sci. USA* **94**, 1634 (1997).
7. E. Knill, I. Chuang, and R. Laflamme, *Phys. Rev. A* **57**, 3348 (1998).
8. G. Feher, *Phys. Rev.* **103**, 834 (1956).
9. Unpublished.
10. H. M. McConnell, C. Heller, T. Cole, and R. W. Fessenden, *J. Am. Chem. Soc.* **82**, 766 (1960).
11. C. H. Bennett, and S. J. Wiesner, *Phys. Rev. Lett.* **69**, 2881 (1992).
12. M. Mehring, J. Mende, and W. Scherer, *Phys. Rev. Lett.* **90**, 153001 (2003).

UTILIZATION OF POLARIZED ELECTRON SPIN OF ORGANIC MOLECULES IN QUANTUM COMPUTING[*]

TIEN-SUNG LIN* AND DAVID J. SLOOP

*Department of Chemistry, Washington University, Campus Box 1134
St. Louis, Missouri 63130, USA*

CHUNG-YUAN MOU

*Department of Chemistry, National Taiwan University
Taipei, Taiwan 106*

The possibility of utilizing highly polarized electron spin of the photo-excited triplet state of organic semiconductors (pentacene molecules) embedded in organic crystals and mesoporous materials by zero-field (ZF) and near zero-field (NZF) pulsed electron paramagnetic resonance (EPR) techniques in a quantum computer will be explored. A simple logic gate, such as CNOT, utilizing such highly polarized electron spins communicating with the surrounding paramagnetic nuclei via hyperfine interaction will be discussed. Major advantages of these approaches are: (1) high electron spin polarization, (2) possible single-molecule detection, (3) orchestrated quantum perturbations can be imposed, and (4) pulsed ZF and NZF EPR techniques can be performed without high magnetic field.

1. Introduction

Quantum computers are potentially superior to classical computers for solving certain classes of computational problems, such as prime factorization, database search algorithms, and quantum mechanical calculations and simulation. For quantum information, the analog of the classical bit of 0 and 1 is the qubit (quantum bit). Qubits are not confined to basic states, say $|0\rangle$ and $|1\rangle$ for a two-level system, but can also exhibit a superposition of basic states, $a|0\rangle + b|1\rangle$. A qubit in such a superposition of states is in both states simultaneously. The basic requirements for building QC are as follows: (1) systems with two-level quantum states, (2) interaction between qubits, and (3) external manipulation of qubits, i.e., QC requires the ability to perform coupled logic, which originated from the couplings in the quantum system involved.[1] Quantum gates can perform multiple logic operations simultaneously in a parallel mode. In a recent

---

[*] Supported partially by NSF (INT0115082) and by PRF grant administered by the American Chemical Society (36970).

205

review, it was stated that one may further utilize a relativistic microchip to create spintronic "phase digits" as the basic unit to encode quantum information.[2]

Several approaches to construct QC have been reported.[1,2] Some powerful quantum search algorithms have been demonstrated with NMR techniques.[3-5] The experimental implementations of the NMR technique in a QC are typically based on two-spin states of a spin-½ nucleus in a magnetic field. Different nuclei in a molecule are distinguishable, so each such spin-½ nucleus can contribute a qubit. The logic gates operating selectively on qubits are implemented by using carefully tailored pulsed rf fields in the solution NMR experiments. For the purpose of performing practical computations, complex logic gates must be implemented, which selectively enable qubit-qubit couplings. Naturally occurring spin-spin interactions can provide such couplings. These NMR-based QC systems rely on quantum logic gates as fundamental elements.

However, the NMR QC suffers from the so-called scalability problem, in which the numbers of molecules in the solution starting in the correct state are given by $\varepsilon 2^{-n}$ (where ε is the Boltzmann population factor, about 10^{-6} for protons at room temperature and 1T, and n is the number of nuclear spins in the molecule). So the molecular quantum interference introduced by molecular complexity will increase dramatically when the intended qubit of a QC increases. Thus, the signal will become exponentially small for complex molecules. Another barrier to practical NMR QC is the decoherence problem. Rotating nuclei in a liquid will begin to lose coherence after an interval of a few seconds to a few minutes as a result of relaxation processes. In NMR, the effective cycle time of a QC is determined by the slowest rate at which the spin order relaxes. This rate is in turn dictated by the interactions between spins and typically ranges from a few to hundreds of cycles per second. However, QC's are running in parallel, which should alleviate the shortcoming of a slow clock cycle. As the size of a molecule increases, the interactions between distant nuclear spins become too weak to be useful for logic gates.

Recently, an extension of NMR to electron paramagnetic resonance (EPR) technique to QC has been reported.[6-8] Though the methodologies of EPR QCs are similar to that of the NMR versions, the greater electron spin magnetic moment improves the Boltzmann factor and therefore the S/N ratio.

2. New Approaches: Polarized Electron Spins in Quantum Computer

The spin Hamiltonian for S = 1 in the presence of a magnetic field is given by the following equation,

$$H = H_d + H_{hf} + g\beta B \cdot S + g_n\beta_n B \cdot I_i \tag{1}$$

where H_d is the electron spin dipolar interaction,

$$H_d = S \cdot T \cdot S = -X S_x^2 - Y S_y^2 - Z S_z^2, \tag{2}$$

The principal axes of the dipolar tensor (T) coincide with the molecular symmetry axes of the organic molecule. X, Y, and Z are the principal values of the dipolar tensor. The eigenstates in zero-field (ZF) are designated as T_x, T_y, and T_z. In the case of pentacene, the long-axis is designated as the x principal axis of the dipolar interaction, the short-axis as y, and out-of-plane axis as z. The H_{hf} term in Eq. (1) is the hyperfine (HFI),

$$H_{hf} = \sum_i S \cdot A_i \cdot I_i = \sum_i S_x A_{ixx} + S_y A_{iyx} + S_z A_{izx})\, I_{ix} + (S_x A_{ixy} + S_y A_{iyy} + S_z A_{izy})\, I_{iy} + (S_x A_{ixz} + S_y A_{iyz} + S_z A_{izz})\, I_{iz} \tag{3}$$

A_i is the hyperfine tensor of the ith nucleus, which consists of both isotropic and anisotropic interactions. The last two terms of Eq. (1) are the electron and nuclear Zeeman terms.

In an effort to overcome the scalability problem in NMR arising from thermalized nuclear spin systems, we explore the possibility of employing the highly polarized electron spin system created in the photo-excited triplet state of pentacene molecules (organic semiconductors, see Fig.1) by laser excitation. This high electron spin polarization (ESP) with ε approaching unity will increase the detectable signal and eliminate the intrinsic limiting factor of thermally populated nuclear spin states in NMR techniques. The manipulation of polarized electron spins to create logic gates can be carried out by pulsed EPR techniques.

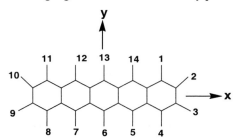

Fig. 1 Molecular symmetry and axes designation of pentacene molecules

In the EPR approach, the electron and nuclear spins are coupled by the HFI, which constitutes the basic qubits for the superposition and entanglement of spin states. The coupling strength at a particular molecular site is proportional to the spin density at that site, which is different from the NMR experiments in which each atom interacts only with a few of its nearest neighbors. In the case of the pentacene molecule, 14 protons can be divided into 4 inequivalent groups because of molecular symmetry. The coupling of three electron spin sub states with four distinct protons will provide 48 two-qubit entangled Bell states. However, the molecular symmetry can be lowered in a partially deuterated sample or in a nitrogen-containing pentacene. Then, in principle, we could have up to 168 two-qubit entangled states using a pentacene-like triplet molecule. The syntheses of functionalized pentacene-like molecules have recently been reported.[9,10] Furthermore, long linear chain dye molecules or polymers with proper chromophores may also be used in QC experiments. Thus, in principle we might find systems to provide higher qubits with high ESP for QC experiments.

We further explore the possibility of embedding the organic semiconductors in nanoscale channels of mesoporous materials, such as MCM-41 solids. The channel diameters of these mesoporous materials can be tailor-made to fit larger molecules: 2 – 30 nm. These mesoporous materials also provide high porosity (ca 1000 m^2/g) which allows for moderate loading of probe molecules with minimal interaction among them, i.e., diluted probe molecule (possibly single-molecule, single-spin) immobilized in the nanoscale channels.   Such an ideal system for QC study has the following attractive features: (1) it reduces the intermolecular interaction and the decoherence interference which will improve the operation cycles, (2) it provides the stability of the organic molecule in which the channel wall behaves as a protective coating, and (3) it preserves the optical integrity allowing the channel wall to be transparent to the laser. Also working with the excited state of a molecule will lift the high purity requirement, unless the impurity affects the optical properties of the system, as a trapping impurity does. The highly spin-polarized state is present only during the preparation of quantum states in photo-excitation. The feasibility of incorporating large organic molecules in the nanochannels of MCM-41 materials has been demonstrated in one of our previous work on C$_{60}$ in modified MCM-41 channel surface.[11] Other groups also reported incorporating copper phthalocyanines and macrocyclic compounds in the nanochannels of MCM-41 materials.[12] We explore the following techniques to test our approaches in QC.

2.1. Pulsed EPR Experiments

The conventional pulsed EPR and electron-nuclear double resonance (ENDOR) experiments can be applied to establish logic gates.[13-15] Below we consider a simple electron + nuclear spins system (a 2-qubit system). The energy level diagram is given in Fig. 2. The quantum states are given as follows,

$$|m_s m_I\rangle = |\uparrow\uparrow\rangle, |\uparrow\downarrow\rangle, |\downarrow\uparrow\rangle, |\downarrow\downarrow\rangle$$

or express them as

$$\varphi_1 = |00\rangle, \; \varphi_2 = |01\rangle, \; \varphi_3 = |10\rangle, \; \varphi_4 = |11\rangle \qquad (4)$$

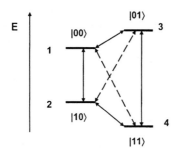

Fig.2 Energy level diagram of two spin system with S = ½ (electron) and I = ½ (nuclear). The solid arrows denote allowed transitions. The dashed arrows denote forbidden transitions (entangled states).

The Bell states of 2-spin ½ (Four possible entangled states) are:

$$\Psi^\pm = [\,|01\rangle \pm |10\rangle\,]/\sqrt{2} \qquad (5)$$

$$\Phi^\pm = [\,|00\rangle \pm |11\rangle\,]/\sqrt{2} \qquad (6)$$

If one starts with the pure state $|10\rangle$, upon application of a selective NMR pulse of $P_I^{34}(\pi/2)$ followed by a selective EPR pulse of $P_S^{24}(\pi)$, we obtain the Ψ^- entangled state.

All other Bell states can be created in a similar way by varying the excitation transition. We note the entangled states carry the phase dependence of corresponding states under rotation about the quantization axis (z-axis), i.e., the unitary transformation, $U_{Sz} = \exp(-i\Phi_1 S_z)$ and $U_{Iz} = \exp(-i\Phi_2 I_z)$ leading to $U_{Sz}U_{Iz}|m_s m_I\rangle = \exp(-i(\Phi_1 m_s + \Phi_2 m_I))|m_s m_I\rangle$.

Fig. 3 Fourier transform of the phase interferograms for phase frequencies: $\nu_1 = 2.0$ MHz and $\nu_2 = 1.5$ MHz; Top (Ψ^-): $\nu_1 - \nu_2 = 0.5$ MHz; Bottom (Φ^+): $\nu_1 + \nu_2 = 3.5$ MHz (From Ref. 8a, courtesy of American Physical Scoiety)

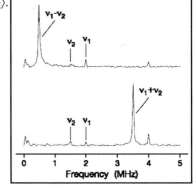

Thus, each of the entangled $\Psi^{\pm}$ and $\Phi^{\pm}$ is characterized by a linear combination $\Phi_1 \pm \Phi_2$ of both phases. Furthermore these states are global states and no local measurement on the single qubits reveals any information about the entangled states.[8] An example is given Fig. 3.

The action of CNOT (controlled-NOT) gate on each of the basis states are given as follows,

$U_{CNOT}|00\rangle = |00\rangle$, $U_{CNOT}|01\rangle = |01\rangle$,
$U_{CNOT}|10\rangle = |11\rangle$, $U_{CNOT}|11\rangle = |10\rangle$.

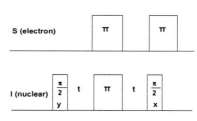

Its matrix representation is

$$U_{CNOT} = \begin{pmatrix} 1 & 0 & 0 & 0 \\ 0 & 1 & 0 & 0 \\ 0 & 0 & 0 & 1 \\ 0 & 0 & 1 & 0 \end{pmatrix} \quad (7)$$

Fig. 4 For the pentacene triplet state, the length of a selective microwave π pulse is ~15 ns. The time delay between two pulses is t ~ 5 ns. A CNOT with about 20 ns and 150 cycles can be performed before decoherence sets in, $T_2 = 3$ μs.

The pulse sequence to execute CNOT operation for such a spin system is shown above in Fig. 4.

2.2. Zero-Field EPR Technique

Recently, we reported a new pulsed EPR technique to study the paramagnetic properties of the photo-excited triplet state of pentacene molecules in zero magnetic fields.[17] The sensitive phase coherence ZF EPR technique allows us to apply proper microwave pulses (for electron spins) and radiofrequency pulses (for nuclear spins) to prepare the superposition of quantum states and to detect the entanglement, so proper logic gates can be constructed. The energy level diagram and the transitions among spin states in ZF and near zero-field (NZF) region are given in Ref. 17. For the sake of presentation, we may consider the pentacene triplet state as a fictitious spin-½ system; because only two spin sublevels will be in resonance during the preparation of superposition. However, the three spin sub states in the triplet state are inter-correlated and complex schemes of preparing superposition and detecting entanglement can be established by careful designs of the preparation of superposition of states and the detection of signal. For instance, we can simultaneously induce both the $T_x - T_z$ and $T_y - T_z$ transitions (T_x, T_y, and T_z are the eigenstates in ZF) and monitor their quantum interference. In ZF or NZF EPR experiments, we can have both the frequency and phase as encoding parameters.

In a selective spin-populated *single-molecule* (pentacene in nanochannels of MCM-41 solids), the isolated electron spin triplet system will interact with nuclei via the HFI. Thus, the molecular complexity should not introduce quantum interference as encountered in NMR experiments. By working with a few spin with very high electron spin polarization (low triplet concentration), instead of an ensemble of spins and capable of selecting a proper organic semiconductor with desired nuclei (such as ^1H, ^2D, ^{14}N, ^{15}N, ^{31}P), we could drastically reduce the scalability and decoherence problems of the NMR QC.

2.3. Non-Adiabatic Passage at Level Anti-Crossing

Previously, we applied a fast-field intensity pulse technique to bring the spin states of the triplet manifold to superposition at the level anti-crossing (LAC) region.[18] If the states are coherently mixed by the fast-field sweeping process, we could examine the entanglement after the LAC passage. From the pulsed ZF EPR measurements of the triplet signal at room temperature, we found that the crossing is a non-adiabatic passage at a field sweeping rate of 2×10^5 T/sec. Detailed examination of the initial population of the photo-excited triplet states indicates that the dominant mechanism for the polarization transfer at the LAC region is the non-adiabatic state mixing by Landau–Zener (LZ) condition.[19] This technique allows us to mix quantum states and examine their entanglement by ZF EPR. We further demonstrated that the rapid field jump at LAC could be examined as an encoded memory in the ZF EPR measurements (see Fig. 5).[18]

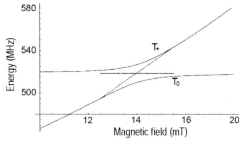

Fig. 5 Energy level diagram of the non-adiabatic passage at level anti-crossing by fast-field sweeping experiments.

When the triplet state is excited at the LAC region, the coherent superposition of quantum states can arise from the off-diagonal terms as a function of the energy gap (γ) given by the following density matrix, where the nonzero off-diagonal terms are responsible for the superposition of substates.

$$\begin{array}{c} |U\rangle \\ |L\rangle \end{array} \left[\begin{array}{cc} \frac{1}{2} & \frac{1}{2}\,e^{-i\gamma t} \\ \frac{1}{2}\,e^{+i\gamma t} & \frac{1}{2} \end{array} \right] \tag{8}$$

where $|U\rangle$ and $|L\rangle$ refer to the upper and lower spin states. Furthermore, each of nuclear spins will also follow the LZ condition. The entanglement can then be examined in the interference of the coherent superposition of the quantum states in the pulsed EPR experiments.

Furthermore, we can adjust the magnetic field sweeping rate to control the composition of state mixing in the level-crossing or level anti-crossing to encode the quantum information. Recently a similar LAC manipulation of Josephson charge states has been applied to control conditional gate operation.[20]

2.4. Pulsed Field Gradient Technique

Previously, we reported a pulsed field gradient (PFG) technique via electron spin echo (ESE) detection for an imaging experiment.[21] The PFG technique is used to create a position-dependent phase shift with a zero average. We thus can apply PFG to encode quantum information. The pulsed EPR technique offers time-resolved information and allows us to investigate the properties and spatial information about paramagnetic species. The field gradients along three orthogonal axes are given by Eq. (9),

$$\begin{aligned} B_Z &= B_0 + (\partial B_Z / \partial z)\Delta z + (\partial B_Z / \partial x)\Delta x + (\partial B_Z / \partial y)\Delta y \\ B_X &= (\partial B_X / \partial x)\Delta x + (\partial B_X / \partial y)\Delta y + (\partial B_X / \partial z)\Delta z \\ B_Y &= (\partial B_Y / \partial x)\Delta x + (\partial B_Y / \partial y)\Delta y + (\partial B_Y / \partial z)\Delta z \\ B &= \left(B_X{}^2 + B_Y{}^2 + B_Z{}^2 \right) \end{aligned} \tag{9}$$

Note that the field gradient technique has also been applied in the NMR quantum computer.[3] The technique produces a spatially varying distribution of states throughout the sample. The PFG will cause the dephasing and reduction of nondiagonal elements of the density matrix. Thus, this method could selectively turn off the transverse (x,y) spin factors in the product operator expansion of the density matrix, while leaving the rest untouched. So the PFG method allows us to spatially prepare a reference spin state or set the system close to ZF conditions by adjusting three mutually orthogonal coils for accurate ZF measurements. Moreover, the phase coherence of our homodyne detection allows us to examine the quantum information in both phase correlation and frequency values.

Acknowledgment

This work was supported partially by a grant from NSF (INT0115082) and a PRF grant administered by the American Chemical Society (36970).

References

1. (a) A. Galindo and M.A. Martin-Delgadot, *Rev. Mod. Phys.* **74**, 347 (2002); (b) C.H. Bennett and D.P. DiVincenzo, *Nature*, **404**, 247 (2000).
2. W. W. Gibbs, *Scientific American*, p57 (September, 2004)
3. (a) N. Gershenfeld and I.L. Chuang, *Science*, **275**, 350 (1998); (b) L.M.K. Vandersypen, M. Steffen, G. Breyta, C.S. Yannoni, R. Cleve, and I.L. Chuang, *Phys. Rev. Lett.* **85**, 5452 (2000).
4. D.G. Corey, A.F. Fahmy, and T.F. Havel, *Proc. Natl. Acad. Sci. USA*, **94**, 1634 (1997).
5. N. Boulant, E.M. Fortunato, M.A.Pravia, G. Teklemariam, D.G. Cory and T.F. Havel, *Phys. Rev. A*, **65**, 024302 (2002).
6. Z.L. Madi, R. Bruschweiler and R.R. Ernst, *J. Chem. Phys.* **109**, 10603 (1998).
7. D. Suter and K. Lim, *Phys. Rev. A*, **65**, 052309 (2002).
8. M. Mehring, J. Mende, and W. Sherer, (a) *Phys. Rev. Lett.* **90**, 153001 (2003); (b) *Phys. Rev. Lett.* **93**, 206603 (2004).
9. J.E. Anthony, J.S. Brooks, D.L. Eaton, and S.R. Parkin, *J. Am. Chem. Soc.* **123**, 9482 (2001).
10. T. Takahashi, M. Kitamura, B. Shen, and K. Nakajima, *J. Am. Chem. Soc.* **122**, 12876 (2000).
11. C.H. Lee, T.S. Lin, and C.Y. Mou, *PhysChemChemPhys.* **4**, 3106 (2002).
12. B.J. Scott, G. Wirnsberger, G.D. Stucky, *Chem. Mater.* **13**, 3140 (2001).
13. T.-S. Lin, *Chem. Rev.* **84**, 1 (1984).
14. D.J. Sloop and T.-S. Lin, *J. Mag. Reson.*, **86**, 156 (1990).
15. A. Schweiger, *Angew. Chem. Int. Ed. Engl.* **30**, 265 (1991).
16. J.-L. Ong, D.J. Sloop and T.-S. Lin, *Chem. Phys. Lett.* **241**, 540 (1995).
17. T.-C. Yang, D.J. Sloop, S.I. Weissman, and T.-S. Lin, *J. Chem. Phys* **113**, 11194 (2000).
18. T.-C. Yang, D.J. Sloop, S.I. Weissman and T.-S. Lin, (a) *Chem. Phys. Lett.* **331**,489 (2000); (b) *Molec. Phys.* **100**, 1333 (2002).
19. (a) L. Landau, *Soviet Phys.*, **1**, 89 (1932); (b) C. Zener, *Proc. Roy. Soc. A*, **137**, 696 (1932).
20. T. Yamamoto, Y.A. Pashkin, O. Astafiev, Y. Nakamura, J.S. Tsai, *Nature* **425**, 941 (2003).
21. D. J. Sloop, H.-L. Yu and T.-S. Lin, *Chem. Phys. Lett.*, **124**, 456 (1986).

QUANTUM SYSTEM IDENTIFICATION[*]

LI-YI HSU[†]

Department of Physics, Chung Yuan Christian University, Chungli, 32023, Taiwan
Republic of China

The joint measurement approach has been shown to detect more information in quantum system. In this paper, we propose a quantum identification algorithm based on the joint quantum measurement. Through this, we can collectively compare an unknown quantum composite system with those in the database. In addition, this study will show that the proposed quantum algorithm can speed up the identification process quadratically.

Suppose there is a database of N quantum composite systems. Each composite system comprises n ($n \ll N$) qubits. The r-th composite system in the database is expressed as

$$\left|X_r\right\rangle = \left|X^1{}_r\right\rangle \otimes \left|X^2{}_r\right\rangle \otimes \cdots \otimes \left|X^n{}_r\right\rangle \tag{1}$$

wherein $\left|X^i{}_r\right\rangle = \cos x^i{}_r \left|0\right\rangle + \sin x^i{}_r \left|1\right\rangle$. Here all the parameters $x^i{}_r$'s are known. In other words, all $x^i{}_r$'s can be regarded as independent and identical random numbers, which are uniformly distributed in the range $-\pi$ to π. Suppose that there is an unknown quantum state, $\left|U\right\rangle$,

$$\left|U\right\rangle = \left|U^1\right\rangle \otimes \left|U^2\right\rangle \otimes \cdots \otimes \left|U^n\right\rangle, \tag{2}$$

where $\left|U^i\right\rangle = \cos u^i{}_r \left|0\right\rangle + \sin u^i{}_r \left|1\right\rangle$. In addition, Alice can access $\left|U\right\rangle$ but she does not know any single value of U^i. On the other hand, she can access the database and knows all $x^i{}_r$'s values. Now, Alice' task is to identify whether or not $\left|U\right\rangle$ is some $\left|X_r\right\rangle$ in the database. Note that $\left|U\right\rangle$ can be some other state than any $\left|X_r\right\rangle$. How can she do it efficiently? The classical analogue

---

[*] This work is financially supported by National Science Council of the Republic of China for under Contract Nos. NSC. 93-2119-M-033 -001.

of this problem is to initially search an unsorted database. That way, Alice can compare two values, X^i_r and U^i, in one step. On the average, it takes about ($N/2$) time steps to find out the answer. Moreover, it takes N times steps to make sure that there is no match in the database. Nevertheless, Alice does not the value of u^i. Using the quantum version of this problem, can Alice speed up the process?

This problem can be regarded as an alternative description of quantum state discrimination [1-4]. Many researches study quantum state discrimination with n being small, and, it is assumed that the unknown state must be the one of the possible states. In our consideration, n can be arbitrarily large. In addition, the unknown state is not necessarily to be one of the possible states. In later discussion, it will be shown that Alice can perform quantum state identification, as previously stated, using joint measurements. As a result, Alice can get correct answer with high confidence after repeating joint measurements about $\sqrt{2^n}$ times.

Here we compare our problem with search problem considered by Grover [5]. In Grover's algorithm on database search problem, since the target (unknown) state is not available for Alice, the oracle operator is needed. In our consideration, some copies of the target state are initially prepared for Alice. In addition, the target state in Grover's algorithm is in some "digital" state, $|\alpha\rangle$, where α is a binary bit string. In our case, all data can be non-digital. In the following we propose how to speed up the identification quadratically based on collective quantum measurements. That is, the query complexity is the same as that of Grover's algorithm. Intuitively, we can compare between $|U\rangle$ and $|X_r\rangle$ qubit by qubit using two-qubit measurements. However, this is equivalent to the comparison of two quantum systems, which does not take any advantage over the comparison of two classical systems [3,4]. Recent studies show that joint measurement can detect more information in the quantum system [6, 7, 8]. In fact, we can compare all $|X_r\rangle$'s to $|U\rangle$ via the joint quantum measurement. Our results showed $O(\sqrt{N})$ joint measurements and thus $O(\ln N \sqrt{N})$ single-qubit measurements are needed to identify $|U\rangle$. Notably, both Grover's algorithm and quantum random-walk search algorithm require $O(\sqrt{N})$ calls to the oracles [5, 9]. In our consideration, $O(\ln N \sqrt{N})$ copies of the unknown state are sufficient to identify the unknown composite system.

Furthermore, since all x^i_r 's are independent, identical and uniformly distributed random variables, all $|X^i_r\rangle$'s were expected to be uniformly distributed in the unit circle lying in the *x-z* plane of the Bloch sphere. In this paper, we presume that $N=2^n$. In other words, there are $|X_1\rangle, |X_2\rangle, ..., |X_N\rangle$ in the database. In addition, $|X_1\rangle, |X_2\rangle, ..., |X_N\rangle$ are required to be linearly independent.

$$\sum_{i=1}^{N} c_i |X_i\rangle = 0 \Rightarrow \forall c_i = 0. \tag{3}$$

In other words, $|X_1\rangle, |X_2\rangle, ..., |X_N\rangle$ span the whole *n*-qubit Hilbert space of dimension N. To be able to make collective identification of some unknown quantum composite system $|U\rangle$, the joint quantum measurement is necessary. We define the one-dimensional projector as

$$\hat{E}_a = p|X_a\rangle\langle X_a|, \tag{4}$$

where p is a positive number and a=1, 2, $\cdots$, N. Clearly, E_a is a positive semi-definite operator. To meet the criteria of quantum measurement, we define

$$\hat{E}_0 = \hat{I} - \sum_{a=1}^{N} \hat{E}_a, \tag{5}$$

where $\hat{I}$ is the identity operator. As a result,

$$\sum_{a=0}^{N} \hat{E}_a = \hat{I}. \tag{6}$$

Suppose that all $\hat{E}_a$, $a = 0,1,\cdots,N$, define the positive operator-valued measure (POVM). As a result, the inequality

$$\langle \psi | \hat{E}_0 | \psi \rangle \geq 0 \ \forall \ |\psi\rangle \tag{7}$$

must be satisfied. In addition, Alice's strategy to identify $|U\rangle$ as follows. If $|U\rangle$ is projected into $\hat{E}_a$ with a being nonzero, it is concluded that $|U\rangle$ could be $|X_a\rangle$. Such conclusion can be correct with probability p. Nevertheless, p is limited under the Eq. (7). Our goal is to approximate the optimal value of p. On the other hand, if $|U\rangle$ is projected into $\hat{E}_0$, the result is inconclusive.

To begin with, note that $|\psi\rangle$ can be written as $\sum_{i=1}^{N} k_i|X_i\rangle$ since $\{|X_i\rangle\}$ span the whole Hilbert space, any n-qubit state. We have

$$\langle\psi|\hat{E}_0|\psi\rangle = 1 - p\sum_{a,r,r'=1}^{N} kk^*\langle X_a|X_r\rangle\langle X_{r'}|X_a\rangle,\tag{8}$$

which comprises four conditions (a) $r=a=r'$ (b) $r=r'\neq a$ (c) $a\neq r\neq r'\neq a$ and (d) either $a=r\neq r'$ or $a=r'\neq r$ for the subscript index r, r' and a. We rewrite $\langle\psi|\hat{E}_0|\psi\rangle$ as follows.

$$\langle\psi|\hat{E}_0|\psi\rangle = 1 - p(T + \sum_{r=1}^{N}|k_r|^2 P_r + \sum_{\substack{r,r'=1\\r\neq r'}}^{N}\mathrm{Re}(k_r k_{r'})S_{rr'} + Y),\tag{9}$$

where

$$T = \sum_{r=1}^{N}|k_r|^2,\tag{10}$$

$$P_r = \sum_{a=1,a\neq r}^{N}|\langle X_a|X_r\rangle|^2,\tag{11}$$

$$S_{rr'} = \sum_{a=1,a\neq r,r'}^{N}\langle X_r|\hat{E}_a|X_{r'}\rangle,\tag{12}$$

and

$$Y = \sum_{\substack{r,r'=1\\r\neq r'}}^{N}\mathrm{Re}(k_r k_{r'})\langle X_r|X_{r'}\rangle.\tag{13}$$

These four terms T, P_r, $S_{rr'}$ and Y are responsible for the conditions of (a), (b), (c) and (d), respectively. For further analysis, we express the expectation values of any function $f(x)$ as $E<f(x)>$. Let X and Y are identical random numbers that are uniformly distributed in the range $-\pi$ to π. Obviously,

$$E < \cos(X - Y) >= 0 \tag{14}$$

$$E < \cos^2(X - Y) >= \frac{1}{2} \tag{15}$$

In addition, $\left| \prod_{i=1}^{n} \cos(x_a^i - x_b^i) \right|$ is evaluated as $\sqrt{E < \prod_{i=1}^{n} \cos^2(x_a^i - x_b^i) >}$.

Therefore,

$$\left| \langle X_a | X_b \rangle \right| = O(\frac{1}{\sqrt{N}}), a \neq b. \tag{16}$$

where $a \neq b$. In other words, any two different systems $|X_a\rangle$ and $|X_b\rangle$ can be regarded as *almost* orthogonal to each other. In other words, $\{|X_i\rangle\}$ can be regarded as the perturbation of some complete set of orthonormal states with order parameter $\frac{1}{\sqrt{N}}$. Therefore, it is easy to verify that T is about $1 \pm O(\sqrt{N^{-1}})$ and the maximal P_r is about $O(1)$. As a result, $\sum_{r=1}^{N} |k_r|^2 P_r$ is about $O(1)$.

Next we consider $S_{rr'}$ in detail. With straight algebra,

$$S_{rr'} = \sum_{a=1, a \neq r, r'}^{N} (\prod_{i=1}^{n} \cos(x^i_a - x^i_r) \cos(x^i_a - x^i_{r'})). \tag{17}$$

Note that $\prod_{i=1}^{n} \cos(x^i_a - x^i_r) \cos(x^i_a - x^i_{r'})$ can be regarded as independent, identical random numbers which are about $O(N^{-1})$ .The maximal Srr' is equal to about the variance of the sum of these random number. Therefore,

$$\max_{r,r'} S_{rr'} = O(\frac{1}{\sqrt{N}}) \tag{18}$$

To evaluate the maximal $\sum_{r,r'=1}^{N} \text{Re}(k_r k_{r'})$, $r \neq r'$, all k_r and k_r are set as

$\frac{1}{\sqrt{N}}(1+O(\frac{1}{\sqrt{N}}))$. Consequently, maximal $\sum_{\substack{r,r'=1 \\ r \neq r'}}^{N} \text{Re}(k_r k_{r'})$ is about $O(N)$.

Therefore, the maximum of $\sum_{r,r'=1, r \neq r}^{N} \text{Re}(k_r k_{r'}) S_{rr'}$ comes close to $O(\sqrt{N})$.

Finally, we evaluate the maximal Y. Obviously,

$$Y \leq \max(\sum_{i=1}^{N} \text{Re}(k_r k_{r'})) \max\langle X_r | X_{r'} \rangle \tag{19}$$

According to Eq. (16), $\max\langle X_r | X_{r'} \rangle$ is about $O(\frac{1}{\sqrt{N}})$. Therefore, maximal

Y is about $O(\sqrt{N})$. Since the logical minimum of $\langle \psi | \hat{E}_0 | \psi \rangle$ is zero. According to Eq. (7), it is required that

$$1 - p(\max(Y) + \max(S_{rr'}) + O(1)) \geq 0 \tag{20}$$

As a result,

$$\max p \leq O(\frac{1}{\sqrt{N}}). \tag{21}$$

In other words, such POVM should be repeatedly performed about $O(\sqrt{N})$ times for identification with high confidence. After one joint measurement is performed, suppose that the unknown state $|U\rangle$ is identified as $|X_r\rangle$ after one collective measurement. For further identifications, any qubit $|U^i\rangle$ should be measured in the basis $\{|X^i_r\rangle, |X^i_r{}^\perp\rangle\}$, where $|X^i_r{}^\perp\rangle = \sin x^i_r |0\rangle - \cos x^i_r |1\rangle$. In the noise-free condition, the post-selection state is always identical to $|X^i_r\rangle$ after the measurement if $|U\rangle$ is really identical to $|X_r\rangle$. Nevertheless, there is another method for further identification. Considering that, we must compare the qubit pair of $\left|U^i\right\rangle$ and $\left|X^i{}_r\right\rangle$ for all $i=1, 2, \cdots, n$. Clearly, for any $\left|X^i{}_r\right\rangle$,

$$
\begin{aligned}
\left|X^i{}_r\right\rangle^{\otimes 2} &= \cos^2 x^i{}_r\left|00\right\rangle + \sin^2 x^i{}_r\left|11\right\rangle \\
&+ \sqrt{2}\sin x^i{}_r \cos x^i{}_r \frac{1}{\sqrt{2}}(\left|01\right\rangle + \left|10\right\rangle)
\end{aligned}
\tag{22}
$$

We can perform a joint measurement on $\left|U^i\right\rangle$ and $\left|X^i{}_r\right\rangle$ in the basis $\{\left|00\right\rangle, \left|11\right\rangle, \sqrt{2}^{-1}(\left|01\right\rangle \pm \left|10\right\rangle)\}$. If $\left|U\right\rangle$ is really identical to $\left|X_r\right\rangle$, after the measurement, the post-selection states are $\left|00\right\rangle$, $\left|11\right\rangle$ and $\sqrt{2}^{-1}(\left|01\right\rangle + \left|10\right\rangle)$ with average probabilities of 3/8, 3/8, 1/4, respectively. It is because all $X^i{}_r$'s are assumed to be identical independent random variables. But, more importantly, the post-selection states can never be $\frac{1}{\sqrt{2}}(\left|01\right\rangle - \left|10\right\rangle)$.

In practice, $\left\langle X_i\middle|X_j\right\rangle$ is likely larger than $O(\sqrt{N^{-1}})$. We can modify the proposed identification as follows. Without loss of generality, suppose that, in our database, for $i=1, 2, \cdots, m$, there is some j such that it is either $\left\langle X_i\middle|X_j\right\rangle > O(\sqrt{N^{-1}})$ or $S_{mj} > O(\sqrt{N^{-1}})$. Recall that all $\left|X_r\right\rangle$'s are required to be linearly independent. That is, $\left|X_1\right\rangle, \left|X_2\right\rangle, \cdots, \left|X_m\right\rangle$ span m dimensions of Hilbert space, H_m. Therefore, we can find m orthonormal states $\left|X_1{}'\right\rangle, \left|X_2{}'\right\rangle, \cdots, \left|X_m{}'\right\rangle$, which also span the same Hilbert space H_m. In addition, $\hat{E}_1, \hat{E}_2, \cdots, \hat{E}_m$ with $\hat{E}'_1, \hat{E}'_2, \cdots, \hat{E}'_m$, where $\hat{E}'_i = p\left|X'_i\right\rangle\left\langle X'_i\right|$, Consequently,

$$
\left\langle X_i'\middle|X_j\right\rangle = S_{ij} = 0,
\tag{23}
$$

where $i=1, 2, \cdots, m$, $j>m$ and . We can regard $\left|X_1{}'\right\rangle, \left|X_2{}'\right\rangle, \cdots, \left|X_m{}'\right\rangle$ as m sets of "nobody" quantum systems. Note that $\left|X_i{}'\right\rangle$'s could be entangled states. Moreover, if $\left|X_i{}'\right\rangle$ is the post-selection state of a collective measurement, such outcome is considered inconclusive. If $m \ll N$, the proposed quantum identification will be much more efficient than classical one.

We, therefore, conclude the proposed quantum identification protocol as follows.

Step 1. Perform and repeat the collective measurement on the unknown quantum states $|U\rangle$ with the measurement basis $\{\hat{E}'_1, \cdots \hat{E}'_m, \hat{E}_{m+1}, \cdots, \hat{E}_N, \cdots, \hat{E}_0\}$.

Step 2. In any joint measurement in the Step 1, if the outcome is some $\hat{E}_r$, where $r \neq 0$ and $r > m$, we hypothesize that $|U\rangle$ is identical to the post-selection state $|X_r\rangle$. If the outcome is $\hat{E}_0$ or some $\hat{E}'_{r'}$, the result is inconclusive.

Step 3. If the outcome is some post-selection state is given by $|X_r\rangle$, further identification should be performed. As previously mentioned, there are two ways of measurements to achieve such task.

Step 4. Repeat the steps 1, 2, and 3 about times to identify the state $|U\rangle$ with high confidence.

Step 5. In the Step 4, if we always fail to identify $|U\rangle$ as $|X_r\rangle$ in Step 3, we can conclude that the unknown state is not identical to any set of quantum systems in the database.

In summary, we investigated on how to identify an unknown composite quantum system. Each qubit state of all composite systems in the database is random distributed in the x-z plane of the Bloch spere. The optimal probability of successful identification in one joint POVM is abut $O(\sqrt{N^{-1}})$. Therefore, the joint POVM should be performed repeatedly about $O(\sqrt{N})$ times to achieve successful identification with high confidence. In addition, further identification is needed once the unknown state is identified as some state in the database. In this way, we can speed up the identification process quadratically.

References

1. I. D. Ivanovic, *Phys. Lett.* **A123**, 257 (1987).
2. A. Peres, *Phys. Lett.* **A128**, 19 (1988).
3. A. Chefles and S. Barnett, *Phys. Lett.* **A250**, 223 (1998).
4. J. N. de Beaudrap, *Phys. Rev.* **A69**, 022307, (2004).
5. L. K. Grover, *Phys. Rev. Lett.* **79**, 325 (1997).
6. A. Peres and P. F. Scudo, *Phys. Rev. Lett.* **87**, 167901 (2001).
7. S. Massar and S. Popescu, *Phys. Rev. Lett.* **74**, 1259 (1995).
8. A. Peres and W. K. Wootters, *Phys. Rev. Lett.* **66**, 1119 (1991).
9. N. Shenvi, J. Kempe, and K. B. Whaley, *Phys. Rev.* **A67**, 052307 (2003).

ORGANIC SEMICONDUCTOR MICRO-PILLAR PROCESSED BY FOCUSED ION BEAM MILLING

HUNG WEN-CHANG, A.ADAWI, A.TAHRAOUI AND A.G.CULLIS

Electronic & Electrical Engineering, University of Sheffield, Mappin Street
Sheffield, S1 3JD, United Kingdom

In order to control light, different strategies have been applied by placing an optically active medium into a semiconductor resonator and certain applications such as LEDs and laser diodes have been commercialized for many years. The possibility of nanoscale optical applications has created great interesting for quantum nanostructure research. Recently, single photon emission has been an active area of quantum dot research. A quantum dot is place between distributed Bragg reflectors (DBRs) within a micro-pillar structure. In this study, we shall report on an active layer composed of an organic material instead of a semiconductor. The micro-pillar structure is fabricated by a focused ion beam (FIB) micro-machining technique. The ultimate target is to achieve a single molecule within the micro-pillar and therefore to enable single photon emission. Here, we demonstrate some results of the fabrication procedure of a 5 micron organic micro-pillar via the focused ion beam and some measurement results from this study. The JEOL 6500 dual column system equipped with both electron and ion beams enables us to observe the fabrication procedure during the milling process. Furthermore, the strategy of the FIB micro-machining method is reported as well.

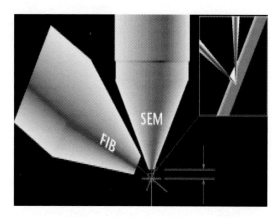

Fig.1 illustration of FIB + SEM dual columns System

223

Introduction

One-dimensional planar microcavities (MCs) have provided a fascinating tool for the study of light-matter interactions [1-2], and also have potential application in optical devices [3]. The MCs structures are the consisting of two mirrors [distributed Bragg reflectors (DBR)] surrounding an optically active region containing a semiconductor layer.

Recently, MCs contains organic semiconductors have been report to operate in the strong coupling regime using thin molecular films within the cavity region [4]. The degree of optical confinement within a MC can be expressed in terms of the cavity Q factor, where $Q=E/\Delta E$ (E is the energy of the cavity mode and ΔE is its linewidth, which is a function of the reflectivity of the cavity mirrors). The interest in creating strong coupled organic MCs is to have large Q factors; therefore, the function of the ΔE is an important consideration. In practice, the function of the ΔE can varied by changing the lateral size of the MCs structures [5], where the micro-pillar MCs structure is we considered in this study.

The convention way of making the micro-pillar is using lateral lithographic patterning [6-7] techniques. However, this techniques required series steps and also leaves an unsmooth surface on the side-wall of the micro-pillar. The side-wall roughness degrades the reflectivity of the DBR mirrors and produces photon leakage; therefore, lower the optical confinement.

Focused ion beam micro-machining techniques [8] provide a simpler way of the micro-pillar fabrication. A 30 keV Ga+ ion beams directly written onto a sample, where the ion dose interact with solid materials, and therefore sputtered materials away. The Figure.2 below shows an illustration of the ion dose interacts with solid materials.

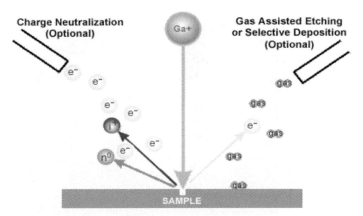

Fig.2. illustration of the ion dose interacts with solid materials

Experiment

Al liquid metal ion source (LMIS) of Ga+ was used with the JEOL 6500F dual columns FIB system. The JEOL6500 system is equipped with both scanning electron microscopy (SEM) and focused ion beam (FIB), which allows observation of the fabrication procedure during the milling. Fig.3 (a) shows the outline structure of the JEOL 6500F system and Fig.3 (b) shows an illustration of the FIB+SEM inside the specimen chamber.

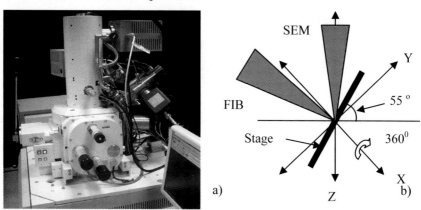

Fig.3 (a) outline structure of JEOL 6500 dual columns system, (b) schematic diagram of FIB+SEM allocation

30kev Ga+ ion energy was applied, while the total emission current was set at 7.2μA, and the probe current was set at about 200pA for milling. The planar organic MCs sample was mounted on a sample block with sample holder, which was 55 degree tilted normal to the ion beam.

The edge of the planar MCs sample was coated with a thin gold layer; this metal layer helps the dissipation of excess charge buildup, and therefore avoids beam shifting during the fabrication.

A circular pattern was designed by Raith software package; Fig.3 (c) shows 20 rings were coupled into a circular shape. The actual FIB scan direction is started from first inner ring towards to outside; this prevents re-deposition effects during the milling process

Scanning electron microscopy (SEM) was used to observe the structures of the cavity, DBR mirrors, and side-wall roughness of the organic micro-pillar MCs samples.

Finally, the MCs fabricated are shown schematically in Fig.3 (d). The bottom DBR was deposited onto a glass substrate by plasma enhanced chemical vapor deposition (PECVD), which was composed of 11 pairs of silicon oxide and silicon nitride. An active polymer film was deposited onto the DBR by spin coating. The top DBR was then deposited on the organic film by thermal evaporation, which was composed of 5 pairs of lithium fluoride and tellurium dioxide.

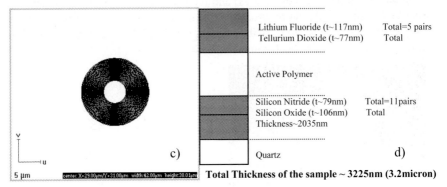

c)

| Lithium Fluoride (t~117nm) | Total=5 pairs |
| Tellurium Dioxide (t~77nm) | Total |

Active Polymer

Silicon Nitride (t~79nm)	Total=11pairs
Silicon Oxide (t~106nm)	Total
Thickness~2035nm	

Quartz

d)

Total Thickness of the sample ~ 3225nm (3.2micron)

Fig. 3 (c) Circular pattern designed by Raith software and, (d) schematic diagram of the organic MCs structures

Results

Fig.4 (a) shows the SEM images of the 5μm micro-pillar MCs fabrication with 30 keV Ga+ ion beam milling for 30mins. 200pA beam probe with 100 single loops scan was milled on the MCs sample; as the result, the side-wall of the micro-pillar structure was very smooth and DBR layers were clear to observe.

The same milling procedure with 4μm micro-pillar MCs structure was shown in Fig.4 (b). The ring pattern was reduced but milling time was the same. As the result, the side-wall of the pillar still was remaining smooth. This implies that we can still reduce the lateral size of the pillar down to approximately 1~2 μm.

Further more, the side-wall roughness was estimated approximately about 50nm under low angle SEM observation.

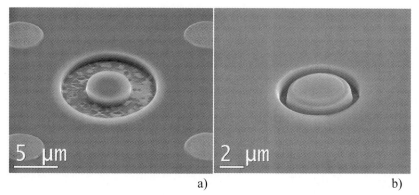

a)   b)

Fig. 4 (a) scanning electron micrograph of organic micro-pillar MCs structures with lateral size of 5μm and, (b) 4μm

Conclusion

Focused ion beam micromachining techniques provide alternative solutions for sub-micron scale fabrication. We demonstrate this method and show the results of smoothing the side-wall roughness of the organic MCs samples. Although we still have to do further reflectivity measurement to micro-pillar sample, we shall improve the milling procedures as to modify the lateral size of the micro-pillar for larger Q factors.

Acknowledgements

The authors would like to thank Alan Walker & Willy Scott (JEOL Europe) for their experimental assistance

Reference

1. M. S. Skolnick, T. A. Fisher, and D. M. Whittaker, Semicond. Sci. Technol. 13, 645 (1998)
2. C. Weisbuch, H. Benisty, and R. Houdre, J. Lumin. 85, 271 (2000)
3. P. K. H. Ho, S. Thomas, R. H. Friend, and N. Tessler, Science 285, 233 (1999)
4. L. G. Connolly, D. G. Linzey, R. Butte, A. M. Adawi, D. M. Whittaker, and M. S. Skolnick, Appl. Phys. Lett. Vol. 83, No. 26, (2003)
5. M. Bayer, A. Forchel, Th. L. Reinecke, P. A. Knipp and S. Rudin, Phys. Stat. Sol. (a) 191, No. 1, 3-32 (2002)
6. J. M. Gerard, D. Barrier, J.Y. Marzin, R. Kuszelewicz, L. Manin, E. Costard, Y. Thierry-Mieg, and T. Rivera, Appl. Phys. Lett. 69, 449 (1996)
7. J. P. Rethmaier, M.Rohner, H. Zull, F. Schafer, A. Forchel, T. L. Reinecke, and P. A. Knipp, Phys. Rev. Lett. 78, 378 (1997)
8. Jon Orloff, Mark Utlaut, and Lynwood Swanson, High Resolution Focused Ion Beams, Kluwer Academic/ Plenum Publishers (2003)

OPTIMAL DESIGN OF SINGLE-PHOTON SOURCE EMISSION FROM A QUANTUM-DOT IN MICRO-PILLAR MICROCAVITY

Y. -L. D. HO, T. CAO, P. S. IVANOV, M. J. CRYAN, I. J. CRADDOCK,
C. J. RAILTON, AND J. G. RARITY

*Centre for Communications Research, Department of Electrical & Electronic
Engineering, University of Bristol
Queen's Building, University Walk, Bristol BS8 1TR, United Kingdom*

We have modelled wavelength scale micro-pillar microcavities of group III-V semiconductor materials using the 3-D finite difference time domain (FDTD) method. A broad band dipole source within the microcavity probes the microcavity mode structure and spectrum. We then investigated the modifications to spontaneous emission of photons form narrowband emitters (e.g. quantum dots) at the centre of the resonance. We find strongly enhanced emission due to small modal volumes and high quality factor (Q-factor). A large fraction of the quantum-dot spontaneous emission is coupled into the fundamental cavity mode. Increasing the number of mirror pairs in the bottom distributed Bragg reflector (DBR) obviously reduces the bottom light leakage, leading to light collection efficiency up to 90%. Moreover, we are now looking at more sophisticated structures with both lateral and perpendicular confinements based on annular and photonic crystal defect cavities in order to suppress the remaining sidewall scattering.

1. Introduction

Secure key exchange has been demonstrated in both optical fibre and free-space [1, 2] by means of the technique of quantum key distribution or quantum cryptography. Present experiments approximate single-photon sources using attenuated laser pulses (standard semiconductor lasers and calibrated attenuators). Thus these are susceptible to attack from eavesdroppers which can select out multi-photon pulses. Hence we aim to design highly efficient and true single-photon emitters in order to eventually replace the attenuated laser. Quantum dots are artificial atoms and they show natural atom-like spectra with narrow excitonic emission lines and can be pumped with short pulse lasers to make single-photon sources [3-5]. Furthermore, Quantum dots have larger quantum confinement energies and can be very efficient light emitters. The high refractive index of group III–V semiconductors means that only a tiny fraction of the emitted light is collected above a typical sample. Light collection efficiency can be increased by using optically pumped micro-pillar microcavities containing single semiconductor quantum dots [6, 7]. These sources can be highly efficient because the high semiconductor refractive index collects a large fraction of the spontaneous emission into the cavity mode. Furthermore the high Q-factor and small modal volumes lead to enhanced spontaneous emission into the mode as first predicted by Purcell [8]. Quasi-three-dimensional (quasi-3D) photonic crystal (PhC) defect microcavities have the smallest possible modal volume while still maintaining high Q-factors [9]. Hence these microcavities are a good candidate for very large Purcell factors. This reduces the exciton lifetime allowing high repetition rate efficient single photon sources to be made.

2. Geometries of Conventional and Annular Micro-pillar Microcavities and Quasi-3D Photonic Crystal Defect Microcavities

Here we will use 3D-FDTD to analyze micro-pillar microcavities [10, 11] and extend our study to more sophisticated structures consisting of quasi-3D photonic defect microcavities with triangular lattices [9]. The micro-pillar microcavity is designed to be made of group III-V semiconductor materials (AlAs/GaAs) with quarter-wavelength-period stacks resonant at the wavelength of 994 nm for large radius. The efficiency of field coupling into the cavity is limited by light leakage from the sidewalls. Hence we look at the structure based on annular [10] and photonic crystal defect cavities [9] formed in the lateral and perpendicular direction in order to suppress side wall leakage (see Figure 1 and Figure 2). In this study we compare three types of annular cavity with 1, 2 and 3 air trench mirrors, micro-pillar microcavities, and PhC defect cavities.

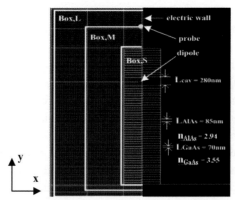

Figure 1. Y-X plane of device showing the mesh, electric wall, probes, quantum dot broadband dipole, cavity thickness (Lcav), and DBR periodicity (LAlAs and LGaAs). Cavity has shown 6 mirror pairs on top and 21 below.

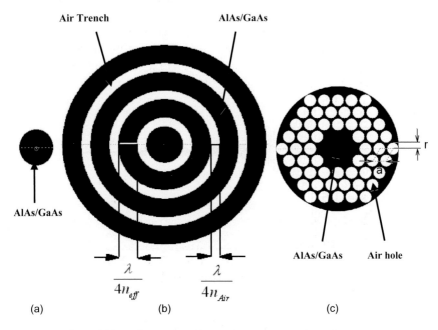

Figure 2. (a), (b), and (c) are the top view of conventional, annular micro-pillar microcavities, and photonic crystal defect microcavity with single quantum dot emitter sitting in the centre of cavity. Black areas represent GaAs/AlAs and white areas air trenches/holes. Pitch a and radius r of the regular air holes are 0.43 μm and 0.19 μm in (c) calculated by MIT Photonic-Bands (MPB) package [12].

3. Simulation Results and Discussions

3.1. *Mode spectrum of micro-pillar microcavity*

We place a broad band dipole source in the centre of the microcavity and input a short few-cycle excitation pulse to probe the mode spectrum. The cavity then rings at its resonant frequency and we monitor the cavity ringdown using a probe above the pillar. We show this for a cavity with 6 top mirror pairs and 21 bottom pairs in Figure 3(a) Taking the Fourier transform of the ringdown signal (in time) allows us to determine the resonant frequencies of the waveguide cavity as shown in Figure 3(b) and also gives us an estimate of Q-factor ($Q=\lambda/\Delta\lambda$).

As we shrink the radius of the micro-pillar microcavity the modal volume (Veff) is reduced which increases the coupling of the dot to the microcavity mode and strongly enhances spontaneous emission (see section 3.4). However the field at

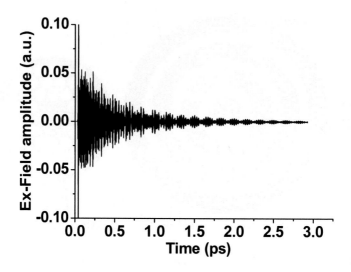

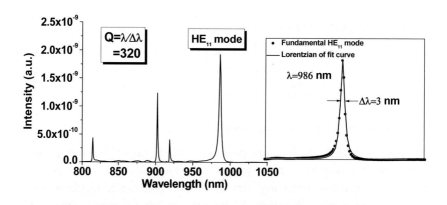

(b)

Figure 3. (a) shows the probed Ex-field amplitude as a function of time for the radius of 0.525 μm of conventional pillar with 6 DBR mirror pairs on top and 21 pairs on bottom. (b) show mode spectrum of conventional pillar. Moreover, the inset of (b) shows the fundamental resonance at blue-shifted wavelength of 986 nm with Q- factor of 320.

the cavity wall becomes stronger, field confinement is reduced and scattering at the interstices between air/GaAs/AlAs becomes stronger. For high Q-factor cavities this leads to a reduction of the Q-factor at small pillar radius (Figure 4). For low Q-factor cavities we see little or no reduction of Q-factor with radius.

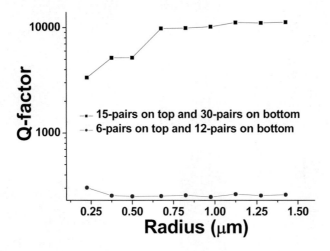

Figure 4. Q-factors of the fundamental HE11 modes of conventional pillar as a function of radius with the structure of 15 DBR pairs on top and 30 pairs on bottom, and the structure of 6 DBR pairs on top and 12 pairs on bottom calculated by the FDTD method.

3.2.   *Electric field amplitude*

Having determined the resonant frequency we can visualise the electric field on-resonance using a single frequency "snapshot". We illustrate the confinement of the electric field amplitude in conventional, annular, and photonic crystal defect cavities in Figure 5.

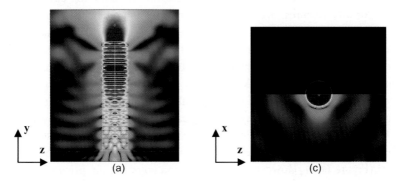

Figure 5. Single frequency snapshot of Ex-field of (a), (b) conventional pillar, (c), (d) annular pillar, and (e), (f) photonic crystal defect microcavity for each plane are in the fundamental HE$_{11}$ mode.

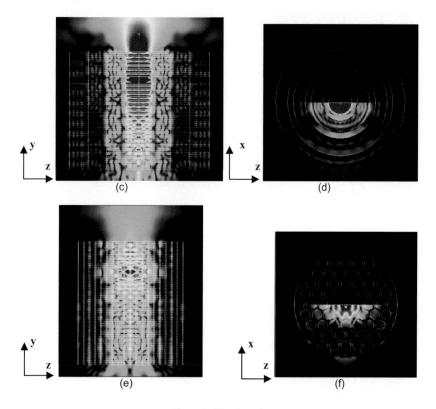

Figure 5. (*Continued*)

From these pictures it is clear that the annular and PhC defect cavities strongly suppress sidewall emission. However emission from below the cavity appears similar for all structures. We have the full information on electric and magnetic fields in our defined snapshot planes. This allows us to calculate the Poynting vector in and thus intensity emitted through planes defined at the sides, top and bottom of the cavities. In Figure 6(a) we compare the fraction of the total intensity lost through side leakage for the different cavity geometries defined above. We plot the fraction of intensity leaking through the bottom mirror as a function of increasing numbers of mirror pairs in Figure 6(b). The quantitative results confirm that the side wall leakage is increasingly suppressed as we increase the number of air trenches and is even more suppressed in the photonic crystal (Figure 6(a)). However this suppression comes at the price of enhanced bottom leakage (Figure 6(b)).

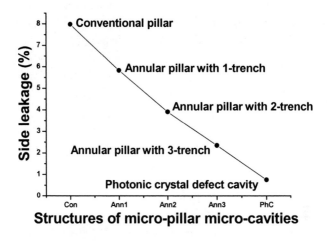

(a)

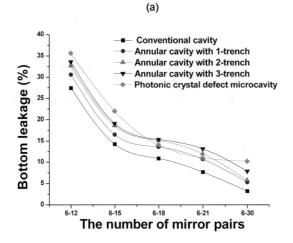

(b)

Figure 6. The side leakage (a) as a function of various cavity geometries with 6-pairs on top and 12–pairs on bottom and the bottom leakage (b) as a function of the number of mirror pairs for the radius of 0.525 μm pillars.

3.3. *Light Collection Efficiency & Spontaneous Emission Enhancement*

Light collection efficiency is estimated from the ratio of the top emitted intensity to the total emitted intensity [5]. In Figure 7 we compare the light collection efficiency for the various geometries. In the case of the simple pillar we see efficiency of more than 90% when we increase the number of bottom mirror pairs. Unfortunately the enhanced bottom leakage in the annular cavities reduces their efficiency to below that of the simple pillar.

$$\eta = \frac{I_{xz}(top)}{I_{total}} \tag{1}$$

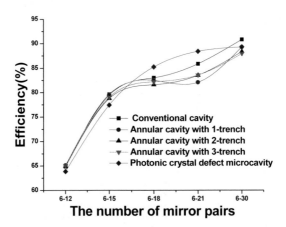

Figure 7. Efficiency estimate from ratio of power emitted from top of device to the total emission as a function of the number of mirror pairs for the radius of 0.525 μ m pillar.

3.4. Purcell Factors

The ratio of total emission from the pillar I_{tot} to total emission in bulk material I_{bulk} should reflect the reduction in spontaneous lifetime induced in a dot. This spontaneous emission enhancement is known as the Purcell factor (F_p).

$$F_p \approx \frac{\tau_{bulk}}{\tau_{cavity}} \approx \frac{I_{total}}{I_{bulk}} \tag{2}$$

A theoretical estimate of Purcell factor can be obtained from [8]

$$F_p \equiv \frac{3Q(\lambda_0 / n_h)^3}{4\pi^2 V_{eff}} \tag{3}$$

Here τ_{cavity} is spontaneous emission life time in the cavity, and τ_{bulk} spontaneous emission life time in the bulk material, V_{eff} effective mode volume and Q cavity quality factor.

We have used our FDTD results to estimate F_p for various pillar radii using a micro-cavity with 6 mirror pairs on top and 12 mirror pairs below and plot the results in Figure 8. We also calculate a Q-factor~260 (see Figure 4) for this cavity and estimate the cavity effective volume to lie between [13]

$$\frac{1}{4}\pi \cdot r^2 \cdot \frac{2\lambda_0}{n_h} < V_{eff} < \frac{1}{3}\pi \cdot r^2 \cdot \frac{2\lambda_0}{n_h}, \qquad (4)$$

where r is the pillar radius.

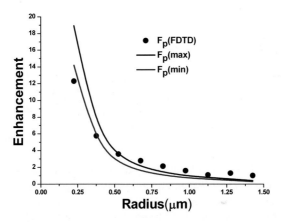

Figure 8. FDTD theoretical estimates of Purcell factor are plotted for the fundamental mode of the micro-pillar micro-cavity as a function of each radius (black dots). Blue and red lines (Fp(max) and Fp(min)) are calculated using Eq. (2) and (3).

The FDTD estimates of Purcell factor are close to predicted values although the simple model (Eq. (2) and (3)) does not hold for large radius ($F_p < 1$ predicted). For small radius derivations are expected because more of the field is in the air region and

$$V_{eff} > \frac{2}{3}\pi r^2 \frac{\lambda_0}{n_h} \qquad (5)$$

4. Conclusions and Further Work

We have shown that more than 90% efficiency emission into fundamental mode is possible with micro-pillar microcavities. Although side leakage is reduced in annular and photonic crystal cavities the bottom leakage is enhanced and efficiencies remain below simple pillar designs. In the future, we will continue the study of various cavity geometries to aim for 99% efficiency emission, perform experiments on real micro-pillars with quantum dots, and study photonic crystal defect microcavities for high-Q operation.

Acknowledgments

We acknowledge funding from EPSRC IRC in Quantum Information Processing (www.qipirc.org), and EU project IST-FET-2001-38864 RAMBOQ (www.ramboq.org).

References

1. C. Kurtsiefer, et al, "Quantum cryptography: a step towards global key distribution," *Nature* **419**, 450 (2002).
2. N. Gisin, et al, "Quantum cryptography," *Rev. Mod. Phys.* 74, 145-195 (2002).
3. J. Kim, O. Benson, H. Kan, and Y. Yamamoto, "A single-photon turnstile device," *Nature* **397**, 500 (1999).
4. P. Michler, et al, "A Quantum Dot Single-Photon Turnstile Device," *Science* **290**, 2282 (2000).
5. J. M. Gérard, et al, "Quantum boxes as active probes for photonic microstructures: The pillar microcavity case," *Appl. Phys. Lett.* **66**, 449-451 (1996).
6. J. M. Ge´rard and B. Gayral,"Strong Purcell effect for InAs quantum boxes in three-dimensional solid-state microcavities," J. Lightwave Technol. **17**, 2089 (1999).
7. M. Pelton, et al, "Three-dimensionally confined modes in micropost microcavities: quality factors and Purcell factors," *IEEE J. of Quantum Electronics* **38**, 170 (2002).
8. E. M. Purcell, "Spontaneous emission probabilities at radio frequencies," *Phys. Rev.*, **69**, 681 (1946).
9. M. Ito, et al, "Enhancement of Cavity-Q in a Quasi-Three- Dimensional Photonic Crystal ,"*Jpn. J. Appl. Phys.* **43**,1990 (2004).
10. Y.-L. D. Ho, et al, "Modelling quantum dots in conventional and annular III-V micro-pillar micro-cavities for single-photon sources," Proceedings of the 4th IEEE Conference on Nanotechnology, Munich, Germany (2004).
11. Y.-L. D. Ho, et al, "Modelling quantum dots in micro-pillar micro-cavities for single-photon sources", International Quantum Electronics Conference, San Francisco, USA (2004).
12. Steven G. Johnson, et al, "Block-iterative frequency-domain methods for Maxwell's equations in a planewave basis," *Optics Express* **8**, 3, 173 (2001).
13. L. C. Andreani, et al, "Strong-coupling regime for quantum boxes in pillar microcavities: Theory," *Phys. Rev. Lett.* **60**, 13 276 (1999).

Professor Maw-Kuen Wu
Minister, National Science Council of
Taiwan

Professor Shan-Hwei Ou
Vice-President of the National Cheng
Kung University

Oleg Astafiev
Recent results in experiments with
Josephson qubits

Toshimasa Fujisawa
Dynamics of single electron charge and
spin in quantum dots

Nobuyuki Imoto
Sending entangled photons:
noisy-channel experiments and
beyond-QKD proposals

Daniel Lidar
Concatenated dynamical decoupling
method against decoherence

Hoi-Kwong Lo
Decoy state quantum key distribution:
The best of both Worlds

Lu Jeu Sham
Quantum information processing and
computing by optical control in a
semiconductor structure

Stephen Bartlett
Decoherence-full subsystems and the
cryptographic power of a private shared
reference frame

Todd Brun
Realistic models of single-spin
measurements via magnetic resonance
force microscopy

Howard Carmichael
Broadband continuous variable
teleportation of a quantum field

Yueh-Nan Chen
W state generation and effect of cavity
photons on the purification of dot-like
single quantum well

Chih Long Chou
Ambiguous and unambiguous
discrimination between quantum states

Chii Dong Chen
Probing charge states in coupled charge
islands

Andrew Doherty
Quantum limits to feedback control of
linear systems

Andrew Dzurak
Silicon qubits using single dopant
atoms

Berthold-Georg Englert
Efficient quantum cryptography with
minimal state tomography

Hsi-Sheng Goan
Device modeling and quantum gate
operations in silicon-based quantum
computing

Masahito Hayashi
Hypothesis testing approach to
quantum information theory

Kurt Jacobs
Limitation on the accessible
information for quantum channels with
inefficient measurements

Jaewan Kim
Quantum imaging

Masato Koashi
Security of quantum key distribution
with strong phase-reference pulse

Dagomir Kaszlikowski
Coherent attacks on quantum key
distribution

Ming-Ting Kuo
Infrared wavelength quantum
communications based on single
electron transistors

Leong Chuan Kwek
A universal quantum network estimator

Keiji Matsumoto
Strong super additivity of entanglement
of formation of stabilizer states

Choo Hiap Oh
Generalized Gisin theorem

Timothy Ralph
Efficient linear optic quantum
computation

Andrew White
Optical quantum computing:
science-fiction, horror-story or news?

Howard Wiseman
Mixed-state entanglement in the light
of pure-state entanglement constrained
by superselection rules

Yen-Fong Chen
Single-photon switching

Robadeh Rahimi Darabad
Pulsed ENDOR-based quantum
information processing

Byoung Ham
Entanglement generation using
coherent standing lights in a condensed
medium

Daniel Y. L. Ho
Optimal design of single-photon source
emission from a quantum dot in
micro-pillar microcavity

Wen-Chang Hung
Organic semiconductor micro-pillar
processed by focused ion beam milling

Jenn Yang Lim
Quantum tomographic cryptography
with Bell diagonal states

Tien-Sung Lin
Utilization of polarized electron spin in
quantum computer

Yin-Zhong Wu
Quantum computation based on
electron spins in quantum dots without
spin-spin interaction

Shigeru Yamashita
Toward a practical environment for
quantum programming

Program

Opening Remarks

Shan-Hwei Ou Vice-President of the National Cheng Kung University
Maw-Kuen Wu Minister, National Science Council of Taiwan

Plenary Talks

Oleg Astafiev Recent results in experiments with Josephson qubits
Toshimasa Fujisawa Dynamics of single electron charge and spin in quantum dots
Nobuyuki Imoto Sending entangled photons: noisy-channel experiments and beyond-QKD proposals
Daniel Lidar Concatenated dynamical decoupling method against decoherence
Hoi-Kwong Lo Decoy state quantum key distribution: The best of both worlds
Lu Jeu Sham Quantum information processing and computing by optical control in a semiconductor structure

Invited Talks

Stephen Bartlett Decoherence-full subsystems and the cryptographic power of a private shared reference frame
Todd Brun Realistic models of single-spin measurements via magnetic resonance force microscopy
Howard Carmichael Broadband continuous variable teleportation of a quantum field
Yueh-Nan Chen W state generation and effect of cavity photons on the purification of dot-like single quantum well
Chih Long Chou Ambiguous and unambiguous discrimination between quantum states
Chii Dong Chen Probing charge states in coupled charge islands
Andrew Doherty Quantum limits to feedback control of linear systems

Andrew Dzurak Silicon qubits using single dopant atoms

Berthold-Georg Englert Efficient quantum cryptography with minimal state tomography

Hsi-Sheng Goan Device modeling and quantum gate operations in silicon-based quantum computing

Masahito Hayashi Hypothesis testing approach to quantum information theory

Kurt Jacobs Limitation on the accessible information for quantum channels with inefficient measurements

Jaewan Kim Quantum imaging

Masato Koashi Security of quantum key distribution with strong phase-reference pulse

Dagomir Kaszlikowski Coherent attacks on quantum key distribution

Ming-Ting Kuo Infrared wavelength quantum communications based on single electron transistors

Leong Chuan Kwek A universal quantum network estimator

Keiji Matsumoto Strong super additivity of entanglement of formation of stabilizer states

Choo Hiap Oh Generalized Gisin theorem

Timothy Ralph Efficient linear optic quantum computation

Andrew White Optical quantum computing: science-fiction, horror-story or news?

Howard Wiseman Mixed-state entanglement in the light of pure-state etanglement cnstrained by superselection rules

Contributed Talks

Yen-Fong Chen Single-photon switching

Robadeh Rahimi Darabad Pulsed ENDOR-based quantum information processing

Byoung Ham Entanglement generation using coherent standing lights in a condensed medium

Daniel Y. L. Ho Optimal design of single-photon source emission from a quantum dot in micro-pillar microcavity

Li-Yi Hsu Quantum identification algorithm

Wen-Chang Hung Organic semiconductor micro-pillar processed by focused ion beam milling

Jenn Yang Lim Quantum tomographic cryptography with Bell diagonal states

Tien-Sung Lin Utilization of polarized electron spin in quantum computer

Yin-Zhong Wu Quantum computation based on electron spins in quantum dots without spin-spin interaction

Shigeru Yamashita Toward a practical environment for quantum
programming

Posters

Ray-Tuan Chang Interactions between two excited Na (3p) atoms
(co-authors: Chin-Chun Tsai, Thou-Jen Whang and
Hui-Wen Wu, Physics Dept., Nat. Cheng Kung U.)

Rei-Hung Chang Information of quantum state
(co-author: Wei-Min Zhang, Physics Dept., Nat. Cheng
Kung U.)

Darwin Gosal Asymmetric multipartite GHZ states and Bell inequalities
(co-authors: D. Kaszlikowski, L. C. Kwek, M. Zukowski and C.
H. Oh, Nat. U., Singapore)

Wen-Chang Hung Organic semiconductor micro-pillar processed by focused
ion beam milling
(co-authors: A.Adawi, A.Tahraoui and A.G.Cullis, U.
Sheffield, UK)

Che-Ming Li Simplification of the encoder-decoder circuit for a perfect
five-qubit error correction
(co-author: Jin-Yuan Hsieh, Nat. Chiao Tung U, Taiwan)

Ying-Yen Liao Orientations of two coupled molecules
(co-authors: Y. N. Chen and D. S. Chuu, Nat. Chiao Tung U.)

Cyrus Lin Wigner rotations, Bell states and Lorentz invariance of
entanglement and von Neumann entropy

Tien-Sheng Lin Quantum routing circuit on the hypercube

Wei-Yang Lin Quantum state diffusion by background hamiltonian
(co-authors: Che-Ming Li, Jin-Yuan Hsieh, Yueh-Nan Chen and
Der-San Chuu, Nat. Chiao-Tung U., Taiwan)

Chuan-Pu Liu Semiconductor quantum dot fabrication and characterization
toward application in single-electron transistors
(co-authors: Hung-Chin Chung, Yen-Lin Lai, Shang-En Wu
and Wen-Huei Chu)

Ru-Fen Liu Stability of entanglement in QED

Guan-Chi Pan Study of the photon switching by quantum interference
(co-authors: Yong-Fan Chen, Yu-Chen Liu, Zen-Hsiang Tsai
and Ite A. Yu, Nat. Tsinghua U., Taiwan)

LIST OF REGISTERED PARTICIPANTS

Astafiev, Oleg RIKEN, NEC, Japan
Bartlett, Stephen U. of Queensland
Brun, Todd U. of Southern California
Carmichael, Howard U. of Auckland
Chang, Chia-Chi Nat. Cheng Kung U., Dept. of Physics
Chang, Rei-Hung Nat. Cheng Kung U.
Chang, Yu-Jen Nat. Cheng Kung U., Institute of
        Micro-Electro-Mechanical-System Engineering
Chang, Li-Kai Tatung U., Dept. of Computer Science Engineering
Chang, Ray-Yuan Nat. Cheng Kung U., Dept. of Physics
Chen, Chia-Chih Nat. Cheng Kung U., Dept. of Physics
Chen, Chii Dong Academia Sinica
Chen, Jiann-yeu Nat. Taiwan U., Dept. of Physics
Chen, Pochung Nat. Tsing Hua U., Dept. of Physics
Chen, Yen-Chih Nat. Cheng Kung U., Dept. of Material Science and
        Engineering
Chen, Chia-Chu Nat. Cheng Kung U.
Chen, Kuan-Ren Nat. Cheng Kung U.
Chen, Yen-ting Nat. Taiwan U., Dept. of Electrical Engineering
Chen, Yi-Chun Nat. Cheng Kung U., Dept. of Physics
Chen, Yong-Fan Nat. Tsing Hua U., Dept. of Physics
Chen, Yueh-Nan Nat. Chiao Tung U., Institute and Dept. of Electrophysics
Cheng, Ching Nat. Cheng Kung U., Dept. of Physics
Cheng, Chuen Ping Nat. Cheng Kung U., Dept. of Chemistry
Cheng, Fu-Chiung Nat. Tatung U.
Cheung, Chi-Yee Academia Sinica
Chiang, Chao-Kai Nat. Tsing Hua U., Dept. of Electrical Engineering
Chin, Chou-hsin Nat. Chiao Tung U., Institute and Dept. of Electrophysics
Chiu, Jian-Lin Nat. Taiwan Normal U., Institute of Electro-Optical Science and
        Techonology
Cho, Ta-Hsiung Nat. Chung Cheng U., Dept. of Physics
Chou, Jui Wen Nat. Chiao Tung U., Institute and Dept. of Electrophysics
Chou, Chih-Lung Chung Yuan Christian U., Dept. of Physics
Chou, Chung-Hsien Institute of Physics, Academia Sinica
Chou, Yao-Hsin Nat. Taiwan U. of Science and Technology
Chu, Wei-chong Nat. Taiwan U., Dept. of Electrical Engineering

Chu, Wen-Huei Nat. Cheng Kung U., Insitite of
        Micro-Electro-Mechanical-System Engineering
Chung, Chun-Jen Nat. Taiwan U.
Chung, Hung-Chin Nat. Cheng Kung U.
Doherty, Andrew U. of Queensland
Englert, Berthold-Georg Nat. U. of Singapore
Fuh, Andy Y.-G. Nat. Cheng Kung U., Dept. of Physics
Fujisawa, Toshimasa NTT Basic Research Laboratories, Japan
Goan, Hsi-Sheng U. of New South Wales, Australia
Gosal, Darwin Nat. U. of Singapore
Guo, Guangcan U. Sc. and Tech, P.R.China
Ham, Byoung Inha U.
Hayashi, Masahito ERATO Quantum Computation and Information Project,
        Japan
He, Jon Hsu Nat. Cheng Kung U., Dept. of Physics
Ho, Michael Southern Taiwan U. of Technology
Ho, Ying-Lung Daniel U. of Bristol, Dept. of Electrical and Electronic
        Engineering
Horng, Shi-Jinn Nat. Taiwan U. of Science and Technology, Dept. of Computer
        Science and Information Engineering
Hsu, Kuang-Yuan MSE, Nat. Cheng Kung U.,
Hsu, Li-Yi Chung Yuan Christian U., Dept. of Physics
Huang, Chin-Hsi Nat. Cheng Kung U.
Huang, Jun-Hang Nat. Cheng Kung U., Dept. of Material Science and
        Engineering
Huang, Shi-Feng Nat. Cheng Kung U., Dept. of Physics
Huang, Hui Feng Nat. Chung Hsing U., Dept. of Physics
Huang, Yi-Kai Nat. Cheng Kung U., Dept. of Material Science and Engineering
Hung, Yu-Ming Chinese Culture U.
Hung, Wen-Chang U. of Sheffield
Imoto, Nobuyuki Osaka U.
Jacobs, Kurt Griffith U.
Ju, Lu Mei Nat. Cheng Kung U., Dept. of Physics
Kang, Chih-Wei Nat. Cheng Kung U., Dept. of Physics
Kaszlikowski, Dagomir Nat. U. of Singapore
Kim, Jaewan Korea Institute for Advanced Study
Ko, Chien-Ming Nat. Taiwan U.
Koashi, Masato Osaka U.
Kuo, Ming-ting Nat. Central U., Dept. of Electrical Engineering
Kuo, Watson Nat. Chung Hsin U., Dept. of Phys
Kwek, Leong Chuan Nat. Inst. Education, Nanyang Tech. U., Singapore
Lai, Yi Feng Nat. Cheng Kung U., Dept. of Materials Science and Engineering
Lee, Boi MSE, Nat. Cheng Kung U.,
Lee, Chien-er Nat. Cheng Kung U., Dept. of Physics
Lee, Ming Tsung Academia Sinica, Institute of Atomic and Molecular Sciences,
Lee, Yuhua Nat. Cheng Kung U., Dept. of Phys.

Li, Ming Shian Nat. Cheng Kung U., Dept. of Physics
Li, Che-Ming Nat. Chiao Tung U., Institute and Dept. of Electrophysics
Li, Chi yen Chung Yuan Christian U., Dept. of Electronic Engineering
Li, Hsiang-nan Academia Sinica, Institute of Physics,
Liao, Ying Yen Nat. Chiao Tung U., Institute and Dept. of Electrophysics
Lidar, Daniel U. of Toronto
Lien, Xanatos Nat. Cheng Kung U., Dept. of Physics
Lim, Jenn Yang Nat. U. of Singapore
Lin, Pe Han Nat. Cheng Kung U., Dept. of Physics
Lin, Yun-Ging Nat. Cheng Kung U., Dept. of Physics
Lin, Cyrus Nat. Cheng Kung U., Dept. of Physics
Lin, Tien-Sheng Lan Yang Institute of Technology
Lin, Tien-Sung Washington U. in St. Louis
Lin, Wei-Yang Nat. Chiao Tung U., Institute and Dept. of Electrophysics
Ling, Alexander Nat. U. of Singapore
Liu, Chuan-Pu Nat. Cheng Kung U., Dept. of Material Science and Engineering
Liu, Ru-Fen Nat. Cheng Kung U.
Lo, Hsin-Pin Chung Yuan Chrsitian U., Dept. of Physics
Lo, Hoi-Kwong U. of Toronto
Looi, Shiang Yong Nat. U. of Singapore
Lu, Hui Wen Nat. Cheng Kung U., Dept. of Physics
Lu, Yan-Ten Nat. Cheng Kung U., Dept. of Physics
Luo, Yuhui Chinese U. of Hong Kong
Matsumoto, Keiji NII, ERATO, Japan
Min, Jen-Fa Nat. Cheng Kung U., Dept. of Physics
Neo, Kai Siang Nat. Cheng Kung U., Dept. of Physics
Ni, Wei-Xin Nat. Nano. Device Lab.
Nyeo, Su-Long Nat. Cheng Kung U.
Oh, Choo Hiap Nat. U. of Singapore
Pan, Guan-Chi Nat. Tsing Hua U., Dept. of Physics
Rahimi Darabad, Robabeh Graduate School of Engineering Science, Osaka U.
Rahman, Taufiqure Independent U., Bangladesh (IUB)
Ralph, Timothy U. of Queensland
Roy, Anirban The Institute of Mathematical Sciences, Chennai, India
Sham, Lu Jeu U. of California San Diego
Soo, Chopin Nat. Cheng Kung U., Dept. of Physics
Sripakdee, Chatchawal King Mongkut's Institute of Technology
 Ladkrabang(KMITL)
Su, Chun Chung Nat. Centr. for High-Performance Computing
Su, Jun-Zhong Nat. Chiao Tung U.
Su, Zheng-Yao Nat. Centr. for High-Performance Computing
Takui, Takeji Graduate School of Sci., Osaka City U.
Tang, Ying Tsan Nat. Chiao Tung U., Institute and Dept. of Electrophysics
Tao, Ming-Hung Nat. Cheng Kung U., Dept. of Computer Science and
 Information Engineering
Tey, Meng Khoon Nat. U. of Singapore

Tsai, Ming-Da Nat. Cheng Kung U., Dept. of Physics
Tsai, Chin-Chun Nat. Cheng Kung U., Dept. of Physics
Tsai, Sientang Southern Taiwan U. of Technology
Wang, Chang-Yi Nat. Tsing Hua U., Dept. of Physics
Wang, Chun-Yen Nat. Cheng Kung U., Dept. of Computer Science and
                 Information Engineering
Wang, Cheng-Yu Nat. Cheng Kung U., Dept. of Material Science and
                 Engineering
Wang, Chuan-Chun Nat. Cheng Kung U., Dept. of Material Science and
                 Engineering
Wang, Kin Weng Nat. Cheng Kung U., Dept. of Material Science and
                 Engineering
Wang, Si-Qi Nat. Taiwan U., Dept. of Electrical Engineering
Wei, Ching-Ming Institute of Physics, Academia Sinica
White, Andrew U. of Queensland
Wiseman, Howard Griffith U.
Wu, Ching Long Nat. Cheng Kung U., Dept. of Physics
Wu, Junde Dept. of Mathematics, Zhejiang U., China
Wu, Meng-Hsiu Nat. Cheng Kung U., Dept. of Physics
Wu, Shin-Tza Nat. Chung Cheng U.
Wu, Yin-Zhong Changshu Institute of Technology and Nat. Cheng Kung U.,
Yamashita, Shigeru Nara Institute of Science and Technology
Yang, Shene-Ming Nat. Cheng Kung U.
Yao, Chao-Chung Nat. Tsing Hua U.
Yen, Cheng-An Nat. Taiwan U. of Science and Technology, Dept. of Computer
                 Science and Information Engineering
Yo, Hwei-Jang Nat. Cheng Kung U., Dept. of Physics
Yu, Hoi-Lai Academia Sinica
Yu, Jui-Ping Nat. Cheng Kung U., Dept. of Physics
Yu, Leo Nat. Chiao Tung U., Elec. Eng. Dept.
Zhang, Wei-Min Nat. Cheng Kung U.
Zhuang, Zhi-Yuang Nat. Cheng Kung U., Dept. of Material Science and
                 Engineering

AUTHOR INDEX